THE TECHNICIAN AS WRITER

Preparing Technical Reports

Ingrid Brunner, Ph.D.
Lehigh County Community College
J. C. Mathes, Ph.D.
University of Michigan
Dwight W. Stevenson, Ph.D.
University of Michigan

BBM Bobbs-Merrill Educational Publishing
Indianapolis

Copyright © 1980 by The Bobbs-Merrill Company, Inc.

Printed in the United States of America

The Bobbs-Merrill Company, Inc.
4300 West 62nd Street
Indianapolis, Indiana 46268

First Edition

Fourth Printing—1983

Cover and interior design by DesignCenter,
Incorporated

**Library of Congress Cataloging in Publication
Data**

Brunner, Ingrid.
    The technician as writer.

    Includes index.
    1. . Technical writing. I. Mathes, John C.,
joint author. II. Stevenson, Dwight W.,
1933-   joint author. III. Title.
T11.B77   808′.066′6021   79-21003
ISBN 0-672-61523-1

Contents

List of Figures

Preface

Technicians, paraprofessionals, and service specialists in all fields today are being asked to do more and more writing. The growing size and complexity of organizations and their increasing need for accurate information means that technicians must write efficiently and effectively. This poses a question: how do technicians learn to write efficiently and effectively, when high school and college courses have not trained them in that kind of writing? Because many college writing courses do not prepare students for the kind of writing they will have to do after graduation, these skills usually must be learned on the job. The purpose of this book is to help you learn these skills in the classroom.

Most technicians, paraprofessionals, and other specialists get their jobs through their skill, experience, and education in a specialized field. We consider a technician to be someone trained in one of various technologies or in such areas as business and commerce, criminal justice, law enforcement, physical therapy, occupational therapy, medical assistance, anesthesiology, public administration, community service, data processing, food services, forestry, or agribusiness. Once employed, technicians find that they must be able to communicate in order to do their jobs and get promoted. One company official states flatly that his company promotes only people who demonstrate the ability to write and speak effectively. In this company, the technician who can produce an effective report soon enjoys higher prestige and rank.

Yet freshman composition and other writing courses often do not train students for these needs. The usual freshman English themes on such topics as "Deception in Advertising," "Energy Conservation," and "The Effect of T.V. Violence on Children" do not prepare the student to write technical reports. The occasional lab reports written to demonstrate students' understanding of experiments do not prepare them to write reports designed for the needs of business organizations.

In most courses, college students write for an audience of one person, a professor; professionals in industry write to a large, varied audience. In college, students write for a reader who is a specialist in the field; in industry, professionals write for many people who are unfamiliar with the field or who know less about the material than the writer. In college, students write to demonstrate mastery of course material; in industry, professionals write to enable people to make decisions and act.

This lack of preparation frustrates the new employee, who firmly believes that the skills and knowledge gained in technical courses back in college adequately prepared him or her for all job needs. Now, on the job, the technician discovers an increasing need for communication skills. A survey conducted by a community college revealed that graduates just starting to work held communications in low esteem in comparison to their technology courses. One year later these same graduates had changed their attitude markedly; they felt that training in communication skills would have contributed much more to their preparation for their work. Painful experience had produced this change.

So what can be done about this problem? What can the technology student and the young working technician do?

We believe the technician must learn to approach the design of technical communications systematically, just as he or she learns to be a competent technician. We believe that if technicians can use systematic procedures to solve technical problems, they can also apply systematic procedures to solve communication problems. The purpose of this book is to present a systematic way to design communications effectively and confidently.

This book concentrates on basic design principles. In that respect, it differs from most other technical writing texts, which often focus on questions of report format, technical style, sentence structure, and mechanics. This book asks questions the writer must answer first: who is to read the report, what do these people want to know, what does the writer want to accomplish, how should the report be constructed to meet these needs? We believe one cannot design a tool, an admissions form, an inventory system, or a report unless one first knows such basic facts as how the product will be used, by whom, and for what purpose.

1.1 Introduction

If a pipe in your bath sprang a leak, you would immediately yell for help to anyone in the house, shut off the valves, and call the plumber. You would give your name, telephone number, address, and the location and extent of the damage, to enable the plumber to fix the leak fast and effectively.

With that telephone call, you would have made a short technical report. Your report would be effective if it enabled the plumber to repair the leak as soon as necessary.

In any office or company, technical reports of all kinds are necessary to get work done. Managers and supervisors need to know what's going on, much as the plumber needs to know about the leak in your bath; other employees need concise and explicit information in order to do their jobs and help you do your job. A good report produces good action.

Oral and written reports are the lifeblood of an organization. Effectively prepared, they contribute to the management and operation of the organization. Ineffectively prepared, they hinder people from doing their jobs well and perhaps affect the profits of a business or the services of an office.

Effective communication of the technician's work is important to his or her company, business, or office. Accordingly, the objectives of this chapter are to help you understand that:
1. As a technician*, you will spend a substantial portion of your time communicating.
2. Technical communication has an instrumental purpose.
3. To prepare a technical report, you first must understand how that report will be used.

---

*As you can see from the list in the preface, we use the single term "technician" for many professions, although for each of you individually there may be a more precise term.

1.2 The Need for Technical Communication

The well-trained, skilled technician with a specialized technical education is a new phenomenon in our society. Before the rapid spread of the community college and the two-year business school, the technician or service specialist was often a high school graduate who received most training right on the job. In many companies there was an immense gap between the untrained technicians, who learned their skills on the job, and the highly trained professionals, many of whom held advanced degrees in business administration, chemistry, electrical engineering, medicine, social work, or other fields.

Today, this gap has been filled by technicians and technologists with a two-year associate's degree or a four-year Bachelor of Technology degree. By the mid-1970s approximately 125,000 students were enrolled in two-year engineering technology programs and 35,000 students were enrolled in Bachelor of Technology programs. In addition, 60,000 students were enrolled in business and commerce technologies, 45,000 in health services and paramedical technologies, 20,000 in public sector curricula, 15,000 in data processing technologies, and about 10,000 in natural science and agricultural technologies. Thousands more were enrolled in certificate programs. The number of these technicians, technologists and service specialists will almost certainly continue to increase throughout the 1980s.

1.2.1 A Case History

Statistics alone cannot explain who this new technician is or what he or she does on the job.

The experiences of one technician illustrate this new role. In 1968, Richard W. Rau was studying for his Associate in Applied Science (A.A.S.) degree in the electronics program at Lehigh County Community College. In April, Bell Labs' personnel recruiter Bill Whelan talked to the students in an electronics class. Mr. Whelan talked further with Richard Rau and two other students after class, and a week later Richard received a letter inviting him for an interview. He then was offered a job with Bell Labs, which he accepted.

After receiving his degree in January, Richard started work for Bell Labs as a technical aide. His job was to assist three members of the technical staff. Three months later, he was assigned to work with Dr. George Marr, whose degree was in nuclear science. Dr. Marr encouraged Richard to acquire increased technical expertise and to write technical reports about their work on ion implantation as a design technology, a breakthrough for Bell Labs.

When Dr. Marr left Bell Labs in 1971, Richard was assigned to Dr. Gilbert Mowery, whose specialty was electrical engineering. Dr. Mowery gave Richard the freedom to accept projects on his own and to write reports under his own name. Richard Rau now works primarily on designs for N-MOS Si-tate L.S.I. custom-designed circuits, reporting directly to Dr. Mowery. He does circuit analysis and physical layout of the chips and generates test vectors to determine whether the chips will function as they should. He also writes and maintains circuit programs, so Bell Labs can guarantee performance.

Richard's success as a technician is obvious. Not so obvious is the fact that Richard's writing skill played a prominent role in his success. He explains, "At Bell Labs, the only way to get ahead was to write." However, he had to develop these writing skills on his own. Although he did well in freshman composition and had a few electronic technology reports at Lehigh County Community College, his real writing education came on the job. He feels that Drs. Marr and Mowery did an excellent job of training him: "They impressed on me the importance of keeping words to a minimum and coming across with the greatest amount of information. I can now sit down and write it with relatively few corrections."

1.2.2 The Importance of Writing Skills

Skill in report writing is the difference between the technician who remains slotted in the lower-level jobs and the one who is promoted. Richard states, "In the real world, whether it is Bell Labs or I.B.M. or any of the other companies, if you do not write, you will not be known. If you want to get ahead, you must be able to write. The technician who cannot write will keep on soldering wires to boards the rest of his career."

Today Richard spends about 20 percent of his time in writing, or "documentation," as he calls it. He submits a brief bimonthly report on his completed projects. You may be surprised to find out that he sometimes has to spend as much as three months on a report. He writes reports for persons in his own division and in other Bell System divisions throughout the country.

In order to write his reports, he has to select from a great mass of information, some of which he has prepared and collected himself in bulky notebooks on each project. Along with a shelf of notebooks, he also has three file drawers completely full of reports and printed materials done by others. In addition, he can visit the excellent Bell library, which contains

all the books and journals any Bell Labs workers might need and also has a complete computer information retrieval system. Thus, he writes his reports not only on the basis of his own work but also according to how well he can digest what others have written.

Richard feels that college pays too much attention to technical skills and not enough to the writing skills necessary on the job. "In college, they keep harping on your major, and the professors are interested only in the technical content of your reports. I wish professors would make sure students learn to put a good report together. You don't have to know everything about math or materials or geography, because you can look these up. I do most of my work on the computer, but I never had a computer course in college. The major obstacle to a technician's success is lack of ability in reading and writing.

"When I got on the job I had to write, write, write, take criticism, and rewrite. You start writing, and you continue writing; the more you do, the better you get—if you keep an open mind and accept criticism. When I send out a report, the phone starts ringing and everybody seems to be criticizing the report. They criticize the spelling and they criticize the numbers. But it is from criticism that you learn. It would have been wonderful if every report I wrote in college had been scrutinized that way."

Richard stresses the need for accuracy in a technical report. He explains that if your writing is technically incorrect, it can hurt you as well as the company. "Once you write it down, it is law: if you're wrong, everyone who uses the information is wrong." He mentions one technician's documentation error which "cost the company $57,000 a second when a circuit failed."

If this makes Richard sound as though he has always loved English, the truth is quite the contrary. Richard says, "I always hated English teachers, and I never thought I'd like writing. But now writing is enjoyable."

Richard expects to be on the professional staff soon, on the same level as employees with a master's degree or a Ph.D. Richard has done very well for himself with an A.A.S. degree.

From Rau's story we can see that technicians usually start by working for persons with greater skill, expertise, and education, relieving them of the more routine, less responsible tasks. This often includes writing, even if the technician never gets his or her name on the finished product. Rau estimates that some 20 percent of the technician's time is spent on the actual writing, and more in oral communication. To keep abreast of the field, he or she also must read books and periodicals in the technical library and the reports arriving on the desk.

1.2.3 A Survey of Technical Report Writers

Of course, you may be in a different field from Richard Rau, and therefore, you may suppose that your work is going to be different. But a survey of all the community college graduates for a three-year period from technical and occupational programs at three Michigan schools (Washtenaw Community College, Ann Arbor; Jackson Community College, Jackson; and Henry Ford Community College, Dearborn) indicates that Richard Rau's experience is typical.[1] The survey includes interviews with representative employees, which support this conclusion. For these employees in various occupational and technical fields, the survey reveals that:

1. The technicians surveyed estimate they spend about half of their time in some kind of communication.
2. Some 95 percent of the employers surveyed estimate that about 30 percent of the technicians' time is spent on oral communication.
3. Technicians worry about the effective organization of their writing; 32 percent have had their writing criticized and 39 percent have encountered poor writing in messages they received.
4. Some 42 percent have to write relatively long reports.
5. Some 42 percent write letters as part of their jobs.
6. Some 29 percent write instructions for higher-ranking employees.
7. Some 39 percent of the industrial technicians and administrative assistants in business write instructions for those outside their own departments.
8. Some 38 percent of the health technicians have to write technical manuals.
9. The therapist and health technicians have to write progress reports and clinical notes evaluating patients' conditions.

The survey further shows that:
1. Employers emphasize the importance of effective oral and written communication.
2. Technicians' opportunities to move upward in the company are largely dependent on their writing skills.
3. The technicians' reports are important criteria for evaluation of job performance.
4. Writing ability is especially important for supervisors, and the chances to be promoted to supervisory positions are limited.
5. Most reports are read by different types of people, most of whom are higher in rank than the writers.
6. Those in health occupations have a great need for reading and speaking skills.

[1] Terry Skelton, "A Survey of On-the-Job Writing Performed by Graduates of Community College Technical and Occupational Programs," *Technical and Professional Writing: Teaching in the Two-Year College, Four-Year College, and Professional School,* Thomas M. Sawyer, editor, Professional Communication Press, Inc., Ann Arbor: 1977. The information includes some additional valuable material from Dr. Skelton's dissertation.

7. Employers understand the importance of skill in reading, because technicians must read specifications, instruction manuals, company rules, trade journals, and product releases. About 62 percent of employers find that further training in reading is often necessary.

Finally, one manager is quoted in the study: "An individual's work is psychologically questioned if his or her writing is sloppy, full of obvious errors, and contains spelling errors in simple, common words. If there are errors in the written presentation, there is an implication of errors in the original work."

You can see that Richard Rau's story is roughly parallel to that of any technical, business, or health service student, and that the key to success as a technician often is the ability to write effectively.

1.3 The Instrumental Purpose of Technical Communication

How can you acquire the ability to write effectively? You must first understand how the piece of writing functions within the company and then just what the writing is intended to achieve.

1.3.1 Keeping to the Specific Purpose of the Report

When you first sit down to write on the job, you probably are attempting a kind of writing completely different from anything you have ever done before. Technical writing is nothing like college assignments, essays, and exams, because each of these was written for one individual, the professor. Nor is technical writing like the writing in a magazine or journal, because that addresses a general audience, the public. Technical writing is written for a specific purpose and addressed to specific audiences within a particular organization. The organization is the company or other institution the technician works for; the purpose is to meet certain specific needs of that organization; and the audiences are the people and groups who must act on the information in the technical writing in order to keep the organization functioning properly. The technical report thus has an instrumental purpose—to get things done.

For instance, Richard Rau works on circuits designed for the individual needs of the customer. He performs circuit analysis and does physical layout of the chips and generates the test vectors and determines whether the chips will function correctly in actual use. His organization is Bell Labs; the purpose of his writing is to explain the latest technical advances to various Bell Labs employees and managers, who use this information in their work.

Because technical writing will be used differently from other writing, it is designed differently from other types of writing. You read an essay or magazine article straight through, expecting to pay equal attention to all parts and to read it top to bottom. Technical writing, however, is seldom designed to be read that way. Since it is used by many readers who are pressed for time and have vastly differing needs and interest, you cannot prepare technical writing the same way you wrote papers in freshman composition class in college. You must learn new methods of designing your writing.

Technical writing is defined and measured by its effect on an organization. Your earlier themes were designed to show logical thinking and organization of ideas on a given topic.

The professor assigned a grade to the paper and did not change his or her behavior because of it. A good or bad paper affected only your grade in the course and had no effect on anyone else. Even your reports in technology and science courses were written only to show the professor your control of the skills and information without affecting anyone else. Technical reports, however, cause changes within the organization's activities. To become effective in technical writing, you must learn to write for distinct, definite purposes. You must learn to write instrumentally.

1.3.2 Your Role in the Organization

To learn the distinct purposes of technical writing, you need to know how technical writing functions in your role as a technician. Your daily activities on the job consist of processing information that comes to you, performing technical and intellectual activities in response to information, and transmitting information to others in your organization who use that information to do their jobs. This process is shown in Fig. 1-1.

Your education has prepared you to do certain things as a technician. To be useful to your company, these technical activities must be changed into information, that is, you must write reports on what you do. This information is what causes ground to be excavated, machinery to be tooled, boxcars to be loaded, sales to increase, and people to live better.

The technician is not a professional technical writer. In many large organizations, some formal reports and articles for persons outside the organization are written by technical writers whose only job is to edit reports and write articles. They do not process information

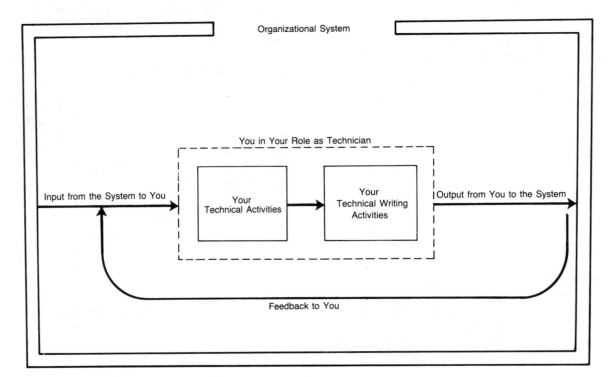

Fig. 1-1
The Technician's Role Related to Company Input
and Output

technically; they merely rewrite and rearrange the information without changing the technical content. In contrast, the technician develops, modifies, and changes information; the technician creates new information for the organization.

Your role in the organization is the most important factor to think of when you do your technical writing. Information, perhaps in the form of an assignment, comes to you. You do certain technical and intellectual tasks. Then you transmit the results to accomplish some purpose related to your role and the assignment. Once it leaves your hands, your technical writing travels through several routes of responsibility and influence inside the organization.

For example, suppose a toy company has a problem. Too many toys are being returned by stores because they are damaged on the shelves. The problem is assigned to a person in the marketing department, who visits several stores, analyzes the situation, and makes recommendations to improve the packaging. These recommendations go into a report to this person's supervisor, who adopts the recommendations. They affect the entire organization, because the supervisor sends the report to the purchasing, production, and claims departments. Other departments also will be affected, shipping, for example. The value of the report is in the various responses and actions it causes in the company; the original analysis of the packaging problem has no actual value in itself.

1.3.3 The Needs of the Readers

When you do technical writing, you must think about the concrete needs of specific people in your company, business, or office, and about the effects your writing will have on the organization. You determine the purpose of your report and design it according to these needs and responses. Technical writing starts in the organization. As a technician, you must see your communication as part of the organization as well as part of your technical activities. The inexperienced technician usually writes only to describe the strictly technical activities, as if he or she were still a student; the experienced technician designs technical writing to serve the needs of the organization.

Exercises

A. Analyze three important technical skills you need to master. Determine the oral and written communication situations required when you use each of these skills on the job. Be prepared to discuss these communication situations with your class.

B. Identify a job skill that requires no communication at all, and prove this to the class.

C. Identify a situation in which you have had to write in order to get something done. Even a personal letter may be an example. How did your awareness of trying to get something done change the way you communicated?

D. Identify a situation in which you have had to write in response to someone else's need. How did responding to a need for information affect the way you communicated?

2 Defining Your Audiences

2.1 Introduction

When you write a letter to your brother, you assume that he won't pass your letter around to other members of the family, especially your mother. You feel free to tell him what you think about your sister's new boyfriend. You know he knows you don't want to tell your parents what you really think of your technical writing course, but you feel free to tell him. And he knows what you're talking about when you mention "family planning."

Suppose, however, that your brother flashes your letter and that your family insists on reading it. Will you be writing to them as you want to? Perhaps you won't be speaking to your sister for a while. You may lose your summer job when your father jokingly mentions your remarks about technical writing to the insurance sales office manager you worked for last summer, who had a new job in mind for you this summer. Finally, what you said about family planning gets you into a great deal of unexpected trouble with your mother.

This is exactly what can happen when you write technical reports without defining your audiences. The technician often assumes he or she is writing a few lines just for an immediate supervisor, who sees the technician every day and is very familiar with the current project. When called upon to report on the project, this supervisor as often as not sends on the technician's hurried lines or incorporates them word for word into a report to fellow employees, the company president, and other managers. Because they are not specialists in the technician's field, these people often are unfamiliar with the project, cannot understand the technical details, and may misinterpret the intentions of the writer. This gets the technician into a great deal of unexpected trouble. Again, serious problems result.

9

People in organizations read reports because they have jobs to do. A report writer's first tasks are to identify the audiences and to determine how the reports will be used. Accordingly, the objectives of this chapter are to enable you to:

1. Understand the need to write and speak to people whose job roles and educational backgrounds differ from your own.
2. Understand how the needs of your audiences affect both what and how you communicate.
3. Identify the specific readers of what you write.
4. Characterize your readers.
5. Classify your readers by how they use what you write.
6. Write a job letter and resume for diverse audiences.

2.2 The Audiences of Your Report

Who reads your reports depends on what kinds of reports you write. The survey of two-year program graduates mentioned in Chapter 1 identifies three basic types of reports: responses to requests for data, responses to inquiries about progress, and recommendations for changes in policy and procedure. Roughly 82 percent of the graduates surveyed were required as part of their jobs to report data of some kind; roughly 50 percent were required to make progress reports; and roughly 42 percent to make recommendations regarding procedures and policy.

2.2.1 Differences Between the Writer and the Readers

The author of the survey cites the case of one technician, Sandy, to illustrate how technicians write to audiences whose roles differ from their own.

Sandy is twenty. She received her associate's degree in accounting in 1975. She has been employed as the head bookkeeper of a local business firm less than one year.

Sandy receives requests for data, requests for reports on assigned projects, and requests for suggestions regarding procedure or policy. She most often communicates data in face-to-face conversations. Her progress on a project is communicated in what she calls "formal reports." Sandy generally sends her suggestions on policy or procedural matters by telephone. She is also required to write instructions for others. These are sent in memos to persons lower in job rank within her department. Sandy writes letters to customers, vendors, and government agencies. The letters are responses to requests for information, letters ordering material or equipment, and letters securing patents and copyrights.[1]

As you can see, Sandy's reports are usually not made to people at her own level in the company or to people trained in her own specialty. Rather, she writes to people at all levels in the company and to people outside the organization, people whose specialties and interests in both cases are often completely different from her own.

2.2.2 Diversity of Readers

To get an idea of the diversity of the audience a technical report writer can face, we need to look at a particular report. Let's consider one written by a technician in a medium-sized computer-oriented service industry. George and his company are real, but, as with many examples in this book, we have changed the names.

Two departments which routinely had to work together were having some trouble getting along. The systems analysis department was putting out work late and therefore having to rerun computer programs. People in the

---

[1] Skelton, pp. 22–23.

programming department apparently did not understand the specific information needs of the systems people. This had been going on for some time, causing some tempers to flare in both the departments.

The technician, George, was a twenty-six-year-old systems analyst with a community college degree in data processing. He was assigned the job of writing a report to solve the situation. He was to explain the specific areas of difficulty, itemize the needs of the systems analysis people, and recommend a way to iron out the current problems. Although George had a handwritten memo from his supervisor, essentially the problem was left to him to resolve—first by analysis, then by communication.

After he had analyzed the department's problem, George had a communication problem. First, look at the distribution of George's audiences around the company. Fig.

2-1 shows the company organization chart. The locations of all George's identified readers are shaded; that is, these recipients were named on the distribution list, although there were others who weren't specifically mentioned. Notice that George's report went not only to members of his own unit, such as his supervisor, but also to units throughout one whole wing of the corporation. Note also that his report went all the way to the front office, to be read by both the president and the executive vice-president.

But Fig. 2-1 does not give you a complete idea of how widely distributed and diverse George's audiences were. Fig. 2-2 identifies each of George's readers by job title, educational background, and primary job responsibilities. Notice that although almost all of George's *primary audiences* (the people who have to act or make decisions on the basis of the report) were in programming, their job training differed substantially. One of them

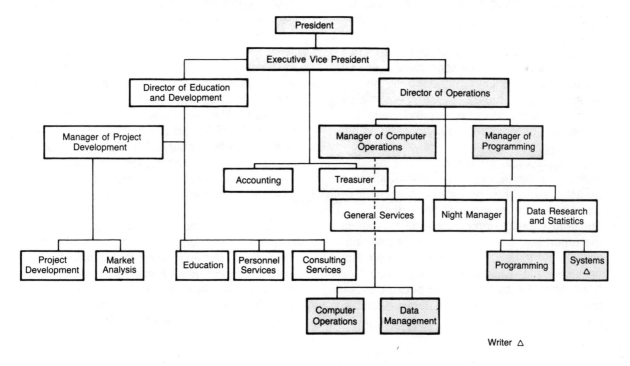

Fig. 2-1
Organization Chart Showing Location of Writer and Readers of One Report

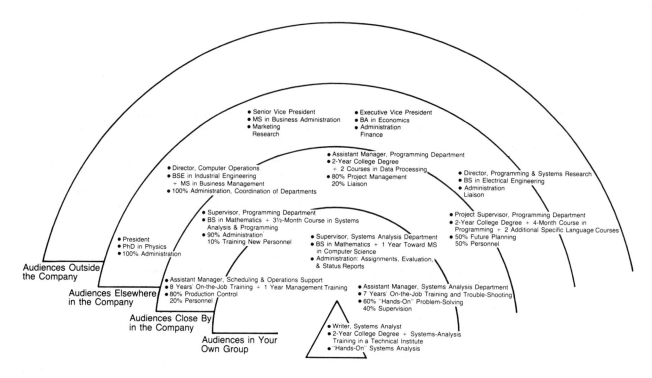

Fig. 2-2
Egocentric Organization Chart for a Systems Department Report

had a B.S. in Math, two had two-year college degrees, and one had on-the-job training. These same people's job needs and responsibilities also differed substantially. One was concerned 90 percent of the time with administration; one was concerned 80 percent with production control; another 50 percent with planning (i.e., long-range planning); and another 60 percent with project direction. Although these readers were in the same wing of the company, their training and their day-to-day needs were clearly not the same.

There were wider audiences for this report, both *secondary audiences* (persons affected by actions the primary audiences would take-in response to the report) and *immediate audiences* (people responsible for evaluating the report and getting it to the right people). Here again the differences were striking. Not only were the readers widely distributed throughout the company, their backgrounds and needs were also quite different. George had to address people with degrees in physics,

economics, electrical engineering, industrial engineering, programming, mathematics, and business administration. Moveover, he had to talk to people with on-the-job training and to people with a Ph.D.

One additional aspect of George's audience isn't apparent. Both Fig. 2-1 and Fig. 2-2 deal only with readers identified on the distribution list, the readers who got the report when it first appeared. Neither shows the unnamed readers who might have had the report handed to them some time afterwards. George's report had to tell these readers, too, what the problem was, what the proposed solutions were, and what they might have to do about it.

To demonstrate how the hidden audiences might be a problem, we asked each of the readers of George's report to estimate how long the report might remain active, that is, how long people might still be reading and using it. Their responses are itemized in Table 2-1. In this case, estimates of the report's lifetime decrease from primary to secondary audiences. At the primary-audience level—in this case, middle and lower management—the report could have a lifetime of from one day to one year. All but one member of the primary audiences believed the report would remain active for at least several months. For these primary readers, the report could remain active until all of its recommendations were in operation and the problem was solved. At the secondary-audience level, on the other hand, the report could have a lifetime of from one day to two weeks. The report has served its purpose for these secondary audiences once they know about it.

2.3 Knowing Your Audiences

As you can see from this example, the technician is responsible for solving problems and interacting with people throughout the company, often at quite different levels. This clearly means that the technician must be able to communicate effectively with people whose training is different from his or her own. Therefore, the technician must have a good understanding of who will read the reports and for what purposes.

In George's case, the report was successful

because it got to the right readers and addressed the right needs. In fact, one of George's readers commented: "This report directly led to a good series of meetings between the two groups, helped establish a clear set of operational guidelines, and opened up channels of communication that are still being used routinely. I think that it is a good report and that the writer should be complimented."

It is not enough for you to realize just in general terms who the audiences are; you must be able to analyze your audiences for every piece of communication you write. In this section, we present a three-step procedure for audience analysis. We have already introduced the concepts; now we make the specific steps of the process clear:
1. Identify specific readers.
2. Characterize the job roles and educational background of those readers.
3. Classify your readers to determine which of them are the most important for your purpose.

Once you master this three-step procedure, you will be able to define your audiences for almost any report you write on the job. In the discussion that follows, we will explain how the steps are performed, one at a time.

Table 2-1
Audience Estimates of One Report's Lifetime

Reader #	Type of Audience	Estimate of Report's Lifetime
1	Immediate	A few days
2	Immediate and Primary	Six months to a year
3	Primary	"Can't estimate; it might be quite some time, because this report sets things in motion."
4	Primary	One month
5	Primary	One day
6	Primary	Three to six months
7	Secondary	A few days
8	Secondary	One day
9	Secondary	Two weeks
10	Secondary	One day
11	Secondary	One day

2.3.1 Identifying Specific Readers

First, you need to know specifically who your readers are. To a large extent, that simply means you should look around the company and ask questions whenever you write. You need to remember that the person who asks you to write—generally, your immediate supervisor—usually does so because he or she needs the information for other people as well as for himself or herself. If you think about it and ask enough questions, usually you can find readers beyond the ones nearest to you in the company. So the first rule, when you are asked to write, is to ask who will read it.

Take the time to do a systematic search for your possible readers. Here, you need to go beyond the immediate audiences, to the audiences even your supervisor may not have thought of. To do this, fill in an *egocentric organization chart*. Write down the names of readers you can identify.

In this kind of chart, unlike the usual pyramidal organization chart (Fig. 2-1), your position does not depend on your job title or how much money you make. You are right in the center, because the routes for your report or memo begin at your desk; you are the center from which the communication radiates.

Now consider the potential audience in the first ring, which represents your own group or department in the company. What specific persons in that group will read or be directly affected by what you write? You need to think beyond the day your report is sent out. Will it go into the file? Will it be distributed later? Will it be combined with other reports for later distribution? To whom? Just remember the example we cited earlier, in which the report was expected to be active as much as a year after it was sent out. As you can see, even in your own group there may well be readers who aren't obvious at first.

Next, consider the second circle, the other groups and departments with whom your group or department frequently interacts. If you are in the Systems Analysis Department, for example, will your reports and memos also be used routinely in the Programming Department or in the Operations Department? If you are in the Testing Department, will your reports be used in Design Department? When you think about it, you will realize that almost any group in a company has "partner groups" with whom it routinely exchanges information and carries out business, and you probably have audiences in these groups as well as in your own.

Now look at the third circle and consider the potential audiences in groups or departments with whom you don't ordinarily interact. Are there any relatively distant management readers? Are there any offices or persons from some fairly remote department of the company involved? Maybe the testing you are reporting is to be used by people in the Publicity Department. Perhaps you are writing something that has to be checked by the Legal Department. Or maybe the parts list you are putting together will have to be priced by Accounting. Again, if you think about it, you may discover some readers you may not even see or talk to very often, but who need what you are working on.

Finally, move to the fourth circle to consider the audience outside the company. These readers might be people who work for some other company or agency, such as a contractor, parts supplier, or government agency. The external audiences might also be people who work for the same company you do, but at a different site. If you consider the long-range possibilities, you may sometimes discover that what you are writing will be sent to people in groups and companies you have never even visited.

When you have finished this first step of your audience analysis, you will have filled in a chart that looks something like that shown in Fig. 2-3. You may have missed a few possible readers. Perhaps you will not have readers in all rings, especially outside the company. However, if you think about your audiences systematically, from those in your own group to those outside the company, and if you ask questions about your audiences, your chances of missing someone important are certainly lessened.

2.3.2 Characterizing Your Readers

To know precisely who your readers are, you need to know more about them than their names. Both job role and educational background, in fact, are really more important

to you than the names, because they tell you something about the kinds of people you are talking to.

To characterize your readers systematically, you need to consider *operational, objective,* and *personal characteristics.* Try to answer and fill in on your egocentric chart the following questions about each of them:
1. What is his or her job title?
2. What are his or her chief responsibilities, concerns, and needs?
3. What is his or her educational and experience background?
4. What personal characteristics should you keep in mind, if any?

Operational characterisics have to do with the person's job role. What does the person do? What does he or she worry about? What are his or her responsibilities? In the example we saw earlier, it was important for George to

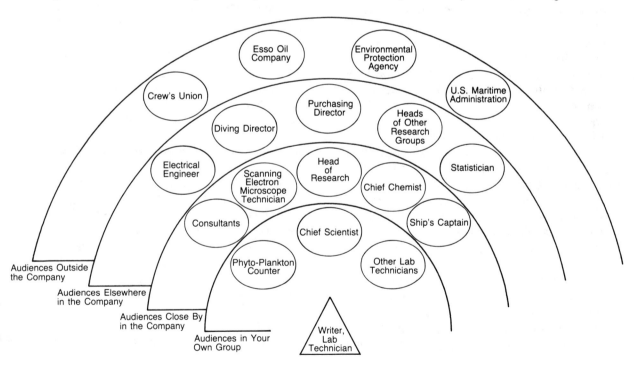

Fig. 2-3
Egocentric Organization Chart for a Test Report on Diving Sampling Techniques

realize that one of his readers, a project supervisor in the Programming Department, was concerned 50 percent of the time with planning and 50 percent of the time with personnel. This fact told George more about the interests of that particular reader of his report than anything else.

Objective characteristics concern the reader's education and experience. It was important for George to understand that among his readers there were many people not trained in systems analysis and programming. There were important readers from a number of different educational disciplines, so it would not have been appropriate for George to write in the technical jargon of his own field. Someone with a business administration or mathematics background would not have understood him at all. As you consider objective characteristics, of course, you should think beyond education. Prior jobs may have given some of your readers a hands-on knowledge of roles other than those they were trained for. In fact, over the years some of your readers may have completely retrained themselves several times. Who your readers are depends on what they know how to do, not just on what they studied in school.

Finally, consider personal characteristics. Although these usually should not pose much of a problem, sometimes they enter in. For example, we know of one manager who likes to skim reports by "looking at the pictures," as he says. He likes lots of tables, graphs, and sketches. Another manager we know can't stand the word "utilize" and is sure to scratch it out and impatiently scribble in "use" any time he comes across it. Still another can't stand the currently common misuse of the word "hopefully," as in "Hopefully, we will have the results by Friday." And yet another manager likes everything to be organized in the form of a numbered list. It doesn't matter what personal wishes or pet peeves your readers may have; it doesn't even matter whether these wishes are logical and sensible. As a writer, you must remember that to talk to somebody the way he or she appreciates, you may have to set your own wishes aside—temporarily, at least. If something matters to your readers, it also matters to you as a writer.

2.3.3 Classifying Your Readers

For you as a writer, the most important aspect of audience analysis is to distinguish your primary and secondary audiences from your immediate audiences. Unfortunately, that isn't always easy to do, and it certainly takes systematic thought on your part. The immediate audiences are usually your project or group leaders, your immediate supervisors. Of course, these people may also be primary or secondary audiences, as well as immediate audiences. But they may be interested in what you write only because they have to transmit it and see that it gets into the right hands. The trick is for you to distinguish which of your readers are the primary audiences and to address the communication primarily to their specific needs. If your immediate supervisor is not one of your primary readers, you need to be aware of that fact and write accordingly.

A convenient way to assign priorities to your audiences is simply to mark on your egocentric organization chart which of the three classes each of your readers falls into. The organization chart in Fig. 2-4 shows how to do this. Notice that this organization chart also identifies each reader and characterizes her or him as discussed above in Steps 1 and 2.

At least at first you should perform this procedure for audience analysis systematically

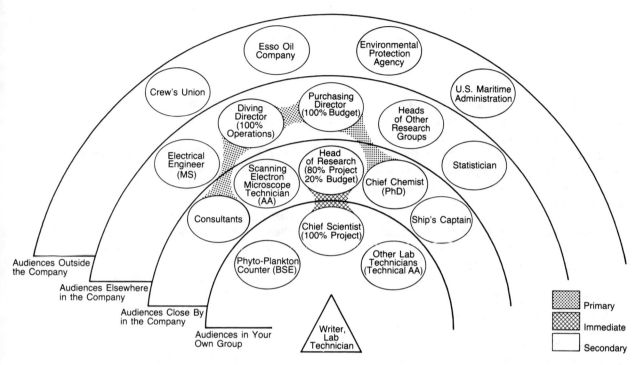

Fig. 2-4
Egocentric Organization Chart for a Test Report on
Diving Sampling Techniques, with Audience Analysis

for each report you write, consciously
identifying, characterizing, and assigning
priorities to your readers. After a while, of
course, when you are accustomed to thinking
of your audiences' needs, the procedure will
become almost automatic. Until that happens,
though, you should be very deliberate about
audience analysis. In fact, you might even start
a file of audience characterizations to help
speed things along when you have reports to
write. Be sure to keep a file of completed
egocentric organization charts for any frequent
or routine types of report. A quick review of
your notes should help avoid forgetting
something important about your readers.
Remember: the needs of your readers
determine what and how you write. Unless you
are thinking systematically about your readers,
you may forget to give them everything they
need.

Exercises

A. Analyze the audiences for a real report and prepare an egocentric organization chart for it. To do this, you are required to interview a report writer, preferably one employed in the field for which you are now training. Use an egocentric organization chart blank (Fig. 2-5) and a writer interview form (Fig. 2-6). Give the completed forms to the instructor, with a one-page synopsis of the communication situation. (Your instructor may also wish you to report your results orally to the class.)

You must follow six steps:
1. Arrange an interview with a person who must write reports and memos as part of his or her job as a technician. In some instances your instructor may prefer to bring that person to class to be interviewed. In this case, all students will complete the assignment on the basis of the interview.

2. Ask the report writer to pick one report or memo that is reasonably important to him or her and to the company.

3. Analyze the audiences for that report or memo, using interview forms (Fig. 2-5 and 2-6) as a means of focusing your questions.

4. Take careful notes during the interviews, especially on the purpose of the report, on the relationships of the report audiences to the writer, on the job roles of those audiences, and on their specific uses for the report. Do not be surprised if the writer hasn't identified or even thought about all of these audiences.

5. Prepare an egocentric organization chart for that report, using the form in Fig. 2-5.

6. Write a one-page synopsis of the interview, explaining the communication situation for the report. You should:

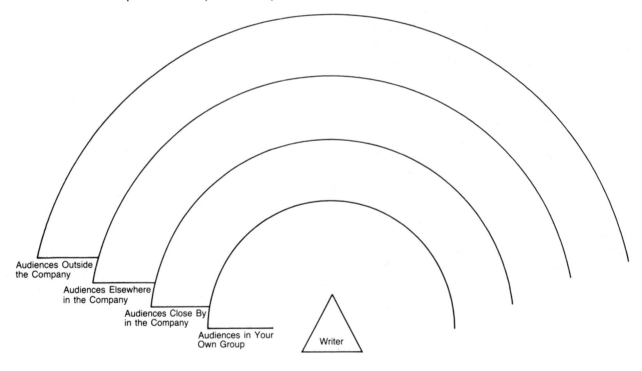

Fig. 2-5
Egocentric Organization Chart Blank

Audiences Outside the Company

Audiences Elsewhere in the Company

Audiences Close By in the Company

Audiences in Your Own Group

Writer

Your name _____

Date   _____

Place of Interview _____

    I. Questions about the writer

        A. Name:
        B. Organization:
        C. Role within the organization:

   II. Questions about the audience

        A. Primary audience (roles and names). The primary
            audience will use the report and act on the
            information it contains.

        B. Secondary audience (roles and names). The secondary
            audience will be directly affected by the report and
            by the organizational uses of the information it
            contains.

        C. Immediate audience (roles and names within the company).
            The immediate audience will transmit the report or the
            information it contains.

  III. Questions about the purpose

        A. What is the conflict or issue of concern to the
            organization?

        B. Does this report respond to previous communication on
            the issue? If so, specifically what were you asked to
            do?

        C. What are the specific technical questions or design
            objectives you have addressed?

        D. Specifically what do you want to accomplish by writing
            this report (i.e., what is your rhetorical purpose)?

Fig. 2-6
Report Writer Interview Form

a. Explain who the writer is.

b. Explain what problem the report addresses and specifically what the report was supposed to accomplish.

c. Explain who the audiences were.

Attach the completed interview form and egocentric organization chart form to your synopsis.

B.   Write a letter applying for a job in your own specialty. Make your application to a specific company and for a specific job opening. Use only factual information about yourself. However, when you write the letter, use the following details as if they were factual:

1. You worked for the company during the previous summer. As a summer employee in the service department, you were responsible for keeping inventory and scheduling reorders. When you left to return to school, your department supervisor, John Harris, said he thought you had done a good job and that he would be happy to have you in his department again. (Note that you are *not* applying for your old job.)

2. You want to work in the department appropriate to your specialty. During your summer at the company, you learned quite a bit about the department and were impressed with it. Although you have not met its supervisor, Will Sexton, you are also impressed by what you have heard about him, and you would like to work under his supervision.

3. The company's personnel office is under the direction of Ann Stenbeck. Although you haven't met Ms. Stenbeck, you remember that when you were hired as a summer employee, you had to go down to her office after your job interview with John Harris to fill out your employment records. Those records are still on file.

Write a job application that will get you an interview at the company and help you to land the job.

3 Stating Your Purpose

3.1 Introduction

There is an old gag letter that circulates around college campuses about once a year. Maybe you have even seen it before. But while the letter tells an old joke, it makes a nice point with which to begin this chapter:

> Dear Mom and Dad,
>
> It has now been three months since I left for college. I have been terrible about writing and I am sorry for my thoughtlessness. Let me bring you up to date now, but before you read on, please sit down. You are not to read any further unless you are sitting down. Okay?
>
> Well, then, I am getting along pretty well now. The skull fracture I got when I jumped out of the window when my room caught fire is pretty well healed. I only spent two weeks in the hospital, and I can see almost normally now and get those sick headaches only once in a while.
>
> Fortunately, my jump was witnessed by an attendant at the gas station nearby, and he was the one who called the Fire Department and the ambulance. He also visited me at the hospital, and since I had nowhere to live after the fire, he invited me to share his apartment. It's really a basement room, but it's kind of nice. He is a fine boy and we have fallen deeply in love and are planning to get married. We haven't set the exact date yet, but it will be before I begin to show.
>
> Yes, Mother and Dad, I am pregnant. I know how much you are looking forward to being grandparents, and I know you will welcome the baby and give it the same love and care you gave me when I was a child. The reason for the delay in our marriage is that my boyfriend has some minor infection which prevents us from passing our blood tests, and I carelessly caught it from him. This will soon clear up with the penicillin injections I am now taking.
>
> In closing, I want to tell you that there was no fire in my room, I did not have a skull fracture. I was

not in the hospital, I am not pregnant, I am not engaged, I do not have syphilis, and there is no man in my life. However, I am getting a D in my history course and a D in French, and I wanted you to see these marks in the proper perspective.

Your loving daughter,
Cindy

P. S. I got an A in psychology.

Yes, Cindy is obviously an A student in psychology. She knew her parents well and could easily predict their normal reaction if she had written the conventional letter home with news about those Ds in history and French. Their reaction would have been of shock and anger. Her purpose in writing the letter, then, was not just to convey the news of her marks in history and French, but also to enable her parents to receive this news without anger. Note that these are two very different purposes. So, instead of writing the conventional letter, Cindy designed a letter to get her parents to accept her bad marks calmly—and maybe even gratefully—assuming that they didn't have heart attacks before they got to the last paragraph.

Although a letter home and a technical report are two completely different sorts of things, Cindy knew how to identify her readers, and she knew precisely what she wanted to accomplish by writing to them. As we saw in the previous chapter, you must be able to identify your readers. But you must also be able to understand and to state what you want to accomplish by writing to those specific readers. If you cannot, you may write a report which is ineffective in design, content, and method of presentation.

The purpose of this chapter is to help you to understand why, how, and where your purpose must be stated in all the technical reports you write. The specific objectives of the chapter are to help you understand:

1. Why you must state your purpose.
2. How the purpose of a report differs from the purpose of your work.
3. How an organizational problem differs from an assignment or technical problem.
4. How to state a purpose for a report effectively and economically.
5. How to write report titles and subject lines which clarify your purpose as well as your topic.
6. Where purpose statements should appear in the technical reports you write.

3.2 Reasons for Stating Your Purpose

Perhaps you can get away without stating your purpose in most of your personal and school writing, such as personal letters, themes, or term papers, but on the job and in your report-writing class, you must precisely state your purpose in every technical report you write. You shouldn't even start writing before you state your purpose. In fact, we can state this axiom: *For every technical report you write, clearly state the specific communication purpose of the report—your reason for writing—both in the title or subject line and in the first paragraph.*

There are three reasons for this:
1. If you do not state your purpose, some of your readers will not understand it.
2. If you do not state your purpose, you may not understand it.
3. If you do not state your purpose, you are being unprofessional.

3.2.1 Readers' Need

You cannot assume all your readers will automatically understand your purpose. As we saw in Chapter 2, almost every report has diverse audiences of readers. Some will read carefully, others hastily. Some will be familiar with your technical area, others will not. Some will read your report because they must in order to do their jobs; others will read it out of a general interest in how they might be affected by what other people are doing. Some will read your report the day it comes out; others will read it months later. Some will know you; others will never have heard of you. Some will know all about the problems you have been working on; others will not.

In short, almost no report addresses audiences all of whom have technical knowledge of your area, an awareness of the problems you are working on, the need to read every word carefully, or the time to pay close attention. Therefore, if you leave your specific purpose unstated, you are gambling that all of your readers will understand why you are writing to them. The odds are that some of your readers will not understand what you want to accomplish unless you tell them. As they say in the army, "There are always 10 percent who don't get the word." That is a rather pessimistic and defensive attitude, but if you want all of your readers to understand your purpose, you need to think defensively. Assume that they will misunderstand what you want to accomplish unless you tell them exactly what it is. State your purpose very openly and explicitly in every report you write.

3.2.2 Writer's Need

If you do not state your purpose, you cannot be sure that you understand it yourself and that your purpose will control how you design and write the report.

It may seem strange that you may not understand your own purpose if you do not

state it. After all, if you do the technical work, why wouldn't you understand the purpose of a report about that work? If you don't understand it, who will? Yet many report writers do not themselves understand why they are writing reports, and this causes a lot of problems.

The main reason report writers sometimes do not understand their own purpose is that they get too caught up in the technical purpose of their work. They forget that the report resulting from their technical work might have a variety of purposes, depending upon who the audiences are. These writers do not distinguish between *technical purposes* and *communication purposes.* You must determine your communication purpose very carefully for every report you write.

The technical purpose does not determine your communication purpose. For example, here is a statement of technical purpose: "The purpose of this investigation was to find an epoxy resin that could be used as a buffing fixture." This clearly defines the technical work the technician was assigned to do. Notice, however, that this statement does not in any sense define the writer's reason for writing a specific report. You could start with this technical purpose and write any number of different types of reports, all with different communication purposes. Here are some possible communication purpose statements:

The purpose of this report is to request an extension of time for the epoxy-resin study.

The purpose of this report is to explain our progress in the epoxy-resin study to date.

The purpose of this report is to provide cost figures to date on the epoxy-resin study.

The purpose of this report is to ask for the assignment of an additional technician to the epoxy-resin study group.

The purpose of this report is to recommend a specific epoxy resin to be used as a buffing fixture.

The purpose of this report is to explain why the epoxy-resin investigation failed and to recommend an alternative investigation.

We could keep going with this one example, but the point should be clear by now. There can be any number of communication purposes for only one technical purpose. To explain the purpose of the report in technical terms is not at all the same as explaining the specific communication purpose.

Getting so engrossed in your technical purpose that you lose sight of your specific communication purpose is an extremely common mistake. If you look at a batch of reports, you will find many which state their purpose, if at all, in technical terms, not communication terms. Here are several additional examples. The annotation in brackets clarifies the real purpose of the reports:

Our purpose was to find whether we have a 7mm ignition wire that will withstand the 350° F and 30 V testing conditions. [Actually the purpose of the report was to state that the company did not have a suitable wire and to recommend that a specific wire be ordered from a supplier.]

Our objective was to calculate the appropriate full-time equivalent work load in the steno pool. [Actually the purpose of the report was to recommend a staffing reduction of eight employees.]

I was assigned to visit the plating department to check on the possibility that there were contaminants getting into the plating bath. [Actually the purpose of the report was to explain a revision in the parts-handling procedure.]

We could go on with examples of purpose statements that suggest the writer did not really understand what the specific purpose of the report was, but the point has been made: state the real purpose of the report, not of the technical work.

The difficulty is that writers frequently lose sight of the fact that every report is a tool designed to accomplish a very specific purpose. They define the purpose for themselves and for their readers in very general, comprehensive, and imprecise terms. They tell what the report is about. This is a natural confusion, but a very troublesome one. It stems from the fact that the technical work is the part most technicians are really interested in. Their communication work tends to get too little serious consideration and is often regarded as just an unpleasant last-minute chore that has to be done. As one of our former students said, "The technical purpose is the fun part of my job; the communication purpose is like cleaning up after the party."

Another difficulty is that sometimes, unfortunately, no one tells the report writers why they were asked to do the technical work. And they don't ask. Or, in some cases, they have been only partially told what their technical work was needed for and therefore have a mistaken notion of the communication purpose. Reports are then written blind, so to speak, by writers who have not been adequately briefed.

Consider this example. A misconduct hearing officer at a large state prison showed us a case report by a correction facilities officer, reporting a prisoner's assault with a handmade knife on another prisoner. The report had been written in such a brief, undetailed, and imprecise fashion that it could not be used as evidence in the misconduct hearing (in-house trial). The result was that the misconduct hearing officer had no choice but to dismiss the charges against the attacker, because he simply did not have enough credible evidence to go on.

This report writer, a new correction facilities officer, did not understand that his report had to be sufficiently detailed and precise enough to be used as evidence in a court case: he did not understand its purpose. He apparently thought that if he just reported the assault in general terms, someone else—the hearing officer, probably—would be responsible for gathering the evidence that would be used in the trial. He just had not been adequately informed about why he had to write a report and what the report had to contain, which should have determined what and how he wrote.

To state your communication purpose, you must first understand it yourself and consider it distinct from your technical purpose. You must take off your technical hat and put on your report writer's hat. That turn from technical work to communication work can make an enormous difference.

3.2.3 Professionalism

If you do not explicitly state your communication purpose in every piece of writing you do, you are not being professional. Technical work is never meant to be a mystery, and technical reports are never supposed to be mystery stories. There should not be any suspense or surprise involved. The technical work should be logical, orderly, and available for anyone to inspect; similarly, the communication work should be right out in the open, with no secrets or trick endings. The trick ending on the letter with which we began this chapter was fine—in a personal letter. On

the other hand, a technical report with a trick ending—or a trick beginning—would be a disaster. Openness and avoiding surprises just go with the territory for technicians, paraprofessionals, and occupational specialists. If you are faithful to your profession, you will state your communication purpose in every report you write. When you do, you help yourself and your readers. You will perform your task best if you know before you ever pick up a pencil why you are doing so. Furthermore, your readers will understand you more clearly when you tell them ahead of time and directly what they are supposed to learn from your reports. The reports you write, like your technical work itself, must be deliberate and open. If you remember that every time you sit down to write, we think you will have an easier time of report writing.

3.3 A Method for Stating Your Purpose

In this section of the chapter, we will explain a simple three-step method for purpose statement. This method can be used as a formula that you can modify for defining your purposes in almost any report. The method consists of stating three things precisely: 1. the problem, 2. the technical tasks you were assigned, and 3. your communication purpose.

3.3.1 The Problem

The communication purpose for a report never exists in a vacuum; it is part of a corporate or organizational activity. For every technical what, there is an organizational why; that is, for everything the technician does, there is a reason. Frequently, indeed most of the time, this organizational context contains a problem and an audience with an interest in it. Your work both as a problem-solving technician and as a report writer is to help the audiences understand the problem and implement solutions. You serve as a kind of bridge between the corporate problems and the corporate audiences interested in them.

You need to know the problem before you can determine the purpose. Thus, your communication purpose cannot be clearly understood or stated unless you first explain the organizational and technical problems at issue. If there were no problems in your organization, there would be no need for reports from you. To state your purpose clearly, begin by pinpointing the corporate problem behind your technical work and your report.

Let's take an example. One technician we know was assigned to find out a great deal of information about the method of manufacture of a particular solvent that his company had been using. That is an assignment of a specific technical task (Step 2 in the three-step purpose statement). But why was the technician asked to do it? What was the reason or problem (Step 1) behind the assignment? Unless we tell you, of course, you cannot guess and therefore you cannot really understand either the assignment or the report stemming from it. That is precisely our point: unless you know the problem, you cannot fully know the writer's purpose. In this example, the problem was that a cheap, easily available commercial solvent which had been used for years by the company for certain testing purposes was going off the market. The company could not find another manufacturer who had a solvent they could use and therefore had to duplicate the formerly available solvent by manufacturing it themselves. To do that, they had to have precise knowledge of its chemical properties and its manufacturing process.

As you can see here, the organizational problem (the old solvent is no longer available) led to an assignment (find the precise chemical properties and method of manufacture of the old solvent). This assignment in turn led to a report used by the readers to develop a solution to the original problem (how can we produce the solvent ourselves?).

You cannot assume your readers already know about the problem behind your technical work. If you consider the nature of the audiences, discussed in Chapter 2, you can see why many of your readers may not now remember the problem that caused you to be assigned the technical work six weeks ago. In fact, some of your readers never knew about the problem. So begin by stating precisely what the organizational problem was.

Here are some examples; as you can see, most can be stated in a few sentences:

Problem: The current flu outbreak coupled with chronic understaffing has almost shut down the print shop. Only two of twelve workers have showed up for the past two days. By Friday two weeks from now (February 20), we have to have the new parts catalogue printed and ready for shipment, but it does not look possible for us to do this in-house with only two or three workers.

Problem: Belleville's present method of sewage treatment removes 75% of the influent phosphate content of the sewage. State law requires removal of 80% of the influent phosphate content.

Problem: Funding agency approval for the parts requisition on contract NAS-27877 has been held up because we did not (as directed) submit an evaluation of electronics parts by an independent laboratory.

Problem: OSHA has cited us for not properly sloping the walls on the excavation at the

Sutherland job site. A fine of $400 has been proposed.

Problem: The drain in the parking lot is apparently plugged. The lot has filled up with water, which is now freezing and thawing, breaking up the pavement.

Problem: Personnel will grant a change of job status from C-3 to C-5 only if we can document the fact that the person in the C-3 position is routinely required to perform C-5 level work.

Problem: The Depot Street bridge has a listed load limit of twelve tons, but the crane that we have to use at the North Campus site and therefore have to move over the bridge is listed at eighteen tons.

Problem: The smoke detector wiring installation proposed for the houses in the Woodview subdivision does not conform to Section 1216.8 of the National Fire Protection Code, because the detectors are not powered "at a point ahead of the main electrical disconnect."

As you can see from these examples, the organizational problem is usually some sort of conflict between what an organization wants to do and what it can or must do. Frequently the problems are people problems, cost problems, or safety problems. Usually they involve factors you would probably not consider technical. Remember, however, that whatever their nature, organizational problems are frequently the direct cause for your assignments, for your technical work, and for your writing reports. For that reason, begin by pinpointing just what the problem is.

We should point out that, in order to understand what the problem is, you may have to go a little out of your way to ask questions. Supervisors often won't take the time to fill you in sufficiently; they just give you an assignment, without telling you why the assignment needs doing. After all, supervisors are as pressed for time as you are and have the same urge to take shortcuts. But don't let that prevent you from asking questions. If you try to find out, at least in general terms, what the problems are, you will probably do a better job of both your technical work and your report writing.

3.3.2 The Assignment

Usually report writers are good about explaining to their readers what they were assigned to do. Writers who have lived the assignment generally don't hold back from telling their readers about it. In a sense, they seem to feel that they are protecting themselves. Sometimes they even get a little carried away with it and overwrite the section of the report that deals with the assignment. In any event, report writers generally do not have to be reminded very strongly to explain their assignments. Therefore, we have only two suggestions.

First, you should identify the source of the assignment when you state the purpose for a report. This is simply a matter of telling your readers who asked you to do what. You say, "On January 15, Mr. Phillips asked me to. . . ." For some reason, report writers often shift into the passive voice and say, "On January 15, I was asked to investigate . . . " or "I was assigned the task of. . . ." This passive construction hides the source of the assignment or request and therefore leaves some useful information unstated. Your readers often need to know who is responsible for your assignment; not to tell them leaves this question unanswered. If yours is the only name mentioned in connection with your report, the responsibility for it is assumed to be yours. Of course, it would be nice to have all the credit, but you probably should share the credit and the responsibility. To make sure that the responsibility is made clear, then, it is a good idea to identify the person who gave you the assignment by name and job title.

Second, you should put your assignment into a form that can be quickly scanned and easily understood. You may not have received it in such neat form, but your readers will appreciate your making things easy for them. To do that, simply state your assignment as a list:

Mr. Harris asked me to do three things:

1. To find the causes of the late deliveries.
2. To determine whether Apex is financially responsible.
3. To recommend a way that we can be certain late deliveries do not happen again.

The list can also be put in the form of questions, for example:

Mr. Harris asked me to answer three questions:

1. Why were the deliveries late?
2. Does Apex have financial responsibility?
3. How can we avoid late deliveries in the future?

One other point about stating your assignments: you may sometimes get very imprecise assignments, because supervisors are always busy and do not always take time to write things clearly or explain them fully. For that reason, a good practice is to put the assignment in writing or clarify the supervisor's writing yourself and then show it to your supervisor for his or her okay before you go ahead with your work. When you come to write the report, you will be more confident that you have done what was intended and that there will be no surprise questions later on.

Here are some further examples of well-stated assignments:

Bill Wolfong asked me to investigate the causes of the increase in failure rate of the airtight seals.

Mr. Munsor instructed me to answer these questions:
1. Does the present system have a maximum safe capacity sufficient to meet our 1990 needs?
2. Are the pumping and storage facilities capable of handling the projected levels of use in 1990?
3. Are the facilities capable of meeting recommended fire-flow needs?

3.3.3 The Communication Purpose

We have already distinguished between your technical purpose and your communication purpose. As we have said, you must state the precise reason you are writing every report; avoid stating only what the report is about or only what your technical purpose has been. State the purpose of your report in a single sentence in the first paragraph. Frequently, as in the examples in Fig. 3-1, you can do this in the last sentence of the first paragraph. However, as in the examples in Fig. 3-2, you can also state your communication purpose in the very first sentence of the report. Either pattern will work, although there is a slight advantage to stating your communication purpose immediately after explaining both the problem and your assignment. If your readers first understand the problem and the assignment, they will be better prepared to understand your reason for writing a report to them.

Use your judgment to determine which pattern is the most effective for each particular report you have to write. A rule of thumb is that your readers should not have to go more than a

The Physical Therapy Department of the University Hospital, Ann Arbor, Michigan, is a participant in a survey conducted by the University Hospital Executive Council. One part of the survey requests the direct patient contact time per man-hour for the department.

The Problem
"They want information; we have no way of giving it to them"

At this time, the department hasn't the means of responding to this question, short of guessing the actual value. My task was to find the direct contact time for the Physical Therapy Department. My purpose for writing this is to propose for your approval a methodology for finding the direct contact time.

Technical Task
"Find the information they need"

Communication Purpose
"I need your approval of this method for getting the information. May I have it?"

The new Feedwater Purity building is due for completion in January 1981 and will involve the addition of several new systems to the plant operations. This will pose a need for plant Operators, Instrument and Control Technicians, Chemistry, and Maintenance personnel to become familiar with and eventually

Problem
Personnel need training, it takes time to train them.

Fig. 3-1
Annotated Examples of Effectively Stated Purposes
Which Begin with the Problem

assume responsibility for applicable systems. To minimize the time spent in familiarizing those involved in the operations, I am proposing a Feedwater Purity Modification Training Session.

Technical Questions/Tasks
How can we save time in training? Perhaps a training session?

The purpose of this report is to explain to you the scope and structure of the proposed training and to request approval for such a training session.

Communication Purpose
"I need your approval for this proposal. May I have it?"

On March 25, 1979, sixteen 8-L semiconductor packages were returned to Murdock Laboratories after failing 80 hours' bias-humidity aging. Upon receiving the packages, I performed ML Probe testing to confirm and determine the mode of failures. The purpose of this report is to present the results of those tests so that Quality Control can be alerted to necessary production modifications.

Problem
"These semiconductors shouldn't be failing, but they are."

Technical Questions
"What caused the failures? Can anything be done about them?"

Communication Purpose
"These results suggest changes in production procedures. Quality control people need to act on that."

Purpose. The addition of the new outpatient wing to the hospital has resulted in an increased amount of paper-

Background Problem
"More paper work is coming our way."

work associated with processing these additional patients. This can be met by adding an operator during the second shift or by upgrading the copying system. A cost analysis indicates that a new copier should be leased.

Problem
Either another operator or an upgraded system could help us cope; we don't know which choice is best.

TQ / Task
(implied) "Which is most cost-effective?"

Communication Purpose
"The copier will work best; I suggest we get one"

Maintenance Department requested that Safety investigate complaints by linemen that our rubber work gloves posed a safety hazard because of ineffective insulation. As part of my summer assignment, I investigated the work situation and the use of the gloves. This report presents the causes of the problem so that an effective safety training program can be implemented.

Problem
"Ineffective insulation is dangerous; our linemen's gloves may be ineffectively insulated"

Technical Questions / Tasks
"What causes the complaints? Is there danger in the situation? Is there danger in the way the gloves are used?"

Communication Purpose
"I propose a training program."

Fig. 3-1
Annotated Examples of Effectively Stated Purposes
Which Begin with the Problem

This report, concerning a 15-year-old state ward, Mark Phillips, is a recommendation for the release of the ward under the community treatment plan. The report represents a consensus of those present at the hearings. The release plan involves the return of Mark to the custody of his parents, full-time enrollment at Morris High School, and an attempt to obtain a part-time job. The release is recommended on the basis of Mark's consistent positive behavior during his stay at Rosedale Training School. Mark is well on his way to rehabilitation, and the risk that he may return to delinquent activity is outweighed by Mark's potential to become a strong, positive young man.

Communication Purpose
"We think Mark should be released."

Technical Questions/Tasks (implied) "Should Mark be released? Under what conditions?"

Problem (implied) "Mark has been in trouble before; there is a risk that even with his training he might get in trouble again."

This is a proposal that a videotape for training new laboratory personnel and work-study students be

Communication Purpose "I propose that we prepare a videotape. Will you approve the project?"

Fig. 3-2
Annotated Examples of Effectively Stated Purposes
Which Begin with the Communication Purpose

prepared to teach them proper handling procedures in the use of radioactive material in all hospital complex laboratories.

Presently there are several laboratories within the hospital complex in which radioactive materials are used. Generally the people involved are adequately trained. However, new laboratory personnel and especially the work-study program personnel are either untrained in the handling of radioactive materials or unfamiliar with the specific procedures in individual laboratories.

Problem

"Safe handling of this stuff requires training; work-study students and new-hires aren't trained."

To illustrate the danger this poses, in one instance observed last week, a work-study student in Laboratory D-7 accidentally exposed himself to low levels of radiation by attempting to clean a radioactive foil used in a gas chromatograph. When asked about his training in the handling of radioactive materials, he indicated that he had none at all. In fact, we later discovered that this work-study

Example of Problem

Fig. 3-2
Annotated Examples of Effectively Stated Purposes
Which Begin with the Communication Purpose

student could not read a dosimeter and could not even define a curie. We believe that on-site training is necessary to orient these new personnel if we are not to risk a serious accident.

Implied Question
Can we train them? The conclusion or answer that leads to the purpose of the report. "Yes we can -- with a videotape."

The purpose of this memo is to recommend purchase of a new torque-testing device.

Communication Purpose
"We need a new device. Will you please purchase it?"

As of October 31, 1979, Marposs will offer a new stepping motor with increased horsepower capabilities. In the past, all stepping motors have undergone bench testing prior to installation. However, due to the increased power of the new system, the existing test equipment is no longer useful. Because the existing equipment cannot be modified to meet the increased demands, a new device must either be purchased or we must forego bench testing.

Problem
"The motors should be tested; we can't test them with existing equipment -- and we can't modify the existing equipment."

Implied Technical Questions or Tasks
Can we modify the equipment? Do we need a new device? Which?

This is a request for authorization of testing alternative ways of applying protective coatings to axles.

Communication Purpose
"I need your authorization. May I have it?"

Fig. 3-2
Annotated Examples of Effectively Stated Purposes
Which Begin with the Communication Purpose

A protective coating applied to the finished axle is required by ACM. Both of the protective coatings now being considered must be applied before the ball joint assemblies are press-fit into the axle, because the high coating-application temperature would destroy the assembly. This results in an economic problem: we must either mask or plug bushing holes to obtain the required press-fit. The purpose of the proposed testing is to determine if the axle can be coated without masking or plugging the bushing holes and still have the ball joint assemblies retain their required press-in and press-out forces.

} *Background Problem*

} *Problem at Hand*
There may be alternatives but, we needed to test them out before we know how effective they are and we can't test them without authorization.

} *Technical Questions or Tasks*
Can the axle be coated without masking or plugging the holes and still be within specs? How?

very few lines into the report before you tell them point-blank why you are writing the report. If you find that for some reason you are delaying the statement of purpose more than that, go back and revise your first paragraph so that it states the purpose immediately.

Whether you put your statement of communication purpose in the first or last sentence of the first paragraph, keep in mind these two suggestions: 1. label the purpose statement clearly, and 2. check the verb or verbs in the statement to assure yourself that you have not confused your communication purpose with your technical purpose.

To label your purpose statement clearly, signalling to your reader that that is what it is, you may use one of two devices: you can put in a heading ("Purpose of this Report") or you can clearly write the signal into the statement of communication purpose ("The purpose of this report is . . ."). Either will draw your readers' attention to the purpose statement. Don't be afraid that you are being too obvious or that repeating the heading or phrase in several reports will become boring. Just remember that if your readers do not understand what your communication purpose is, you have wasted both their time and your own. The greater likelihood that your report will be effective if your purpose is clear to your readers offsets the repetitiveness of your pattern for the purpose statement. Purpose statements are too important to be manipulated just for the sake of variety.

To assure that you have not inadvertently confused your communication purpose with your technical purpose, just check the verbs in the purpose statement. Ordinarily you will use a pattern like this: "The purpose of this report is to [verb]." When you check this slot, you should find verbs which signal communication,

such as *ask, recommend, inquire, request, propose, authorize, respond,* etc. Go back and revise the statement if you find verbs which signal technical activity, such as *find, test, investigate, examine, identify, solve.* All of these verbs signal technical activity, not communication activity. They define the purpose of your technical work, but not the purpose of your communication resulting from that work.

For example, you might say the technical purpose of an investigation was "to compare two packaging systems"; the communication purpose resulting from that investigation would be "to recommend a packaging system." As another example, if the technical purpose is "to analyze a soil sample," the communication purpose resulting from that analysis might be "to advise you that the soil sample is deficient in phosphate." The verbs you use in your purpose statement are signals of your thinking. If you have changed from your technical hat to your communication hat, your verbs will signal communication activity. If you are still wearing your technical hat, your verbs will signal technical activity.

Communication purpose statements should always make clear what you want to happen. If you want approval for a proposal, say so. If you want to ask for help, say so. If you want to recommend an action, say so. One manager we know often asks report writers, "Okay, I've read your report, now what do you want me to do?" Your purpose statement should make that clear. Ask yourself the question "Why?" after you write the purpose statement. For example, "The purpose of this report is to summarize our findings." "Why?" Perhaps the findings suggest that something is really wrong and that more time is needed to check this out. If so, the purpose of the report is not just to summarize the findings; it is to summarize the preliminary findings and ask for authorization of continued testing. Ask yourself what you want to happen. If you do that, you can be very explicit about your real reason for writing. Once you have clarified this for yourself, make it clear to your readers.

To summarize, when you state the purpose of the report, be sure that you have made three concepts clear:
1. The problem.
2. Your assignment.
3. The communication purpose of your report.

If you routinely make all three of these concepts clear and place and label the statement well, your readers should be able to understand your purpose for sending them a report. To help you recognize effective statements of purpose from reports, those in Fig. 3-3, are annotated to indicate the various features we have discussed.

3.3.4 Abbreviated Communication Purposes

We need to consider one more point about purpose statements. Do all purpose statements need the three elements we have just defined—problem, assignment, and communication purpose? Aren't there times when you can take a shortcut, abbreviating the purpose statement? The simple formula for the purpose statement will work perfectly in most of the reports you write, but there are times when you can abbreviate the statement by omitting the explanation of the problem, the assignment, or both:
1. When the audience for your report is so limited that all your readers know about the problem and assignment.
2. When your report is a routine report of progress occasioned not by an immediate problem but by a set reporting date.
3. When your report is intended not so much

to cause a specific response as to provide documentation for the file.

In all three of these cases, you can usually compress your purpose statement somewhat; if you are careful, you may even be able to eliminate both the problem and the assignment.

Suppose you are working on a project which requires periodic progress reports—one every two weeks, for instance. In such a case, it is probably sufficient to say something like: "This report explains our progress on the project to reforest the 450 acres of recently burned-off land in McLean County." That statement includes a little of the problem (burned-off land) but focuses on the purpose (progress report). Suppose that you are to file a report on expenses for the record. You might say, "The purpose of this report is to itemize the expenses of our November 10th site visit to Austin, Texas." The purpose is clear, but the writer has chosen not to explain the reason for the site visit or the standard request for reports of expenses.

Such reports have extremely limited audiences, who can be expected to know about a problem and will therefore need only a brief reminder. Reports often are written for no other reason than to comply with a reporting schedule. Reports also are written for the file. If you are certain that the problem and your assignment are familiar to all your readers and that there is no immediate problem, you can shorten your purpose statement to a statement of your communication purpose.

Remember that you should simplify your purpose statement like this only in situations that permit it. Think of your primary and secondary audiences and think of how the report might be used at a later date. If there

According to the Pontiac Building Code, Spencer Construction must install one smoke detector in our Ponderosa Ranch styles and two in our Sierra View styles. D. Finton, the Project Engineer, has chosen the Insta-Alarum 23 photoelectric smoke detector for its reliability and the low bid from the manufacturer's representative. I was asked to describe this detector so that our sales agents can assure prospective clients that Spencer Construction has installed a reliable detector that meets code specifications.

Background Problem
To meet code we must install a detector; we don't know which is best.

Solution to Background Problem
"We have chosen one to install."

New Problem and Communication Purpose
"Sales agents need to know about our choice but don't. This document tell them all about it."

At your request, I investigated Mini Wear Types I and II carbon paper for quality and efficiency. I tested for durability, multi-copy legibility, curling tendency, and erasability and found Type I carbon paper to be superior to Type II. I recommend that Mini Wear Type I carbon paper be used in Central Typing.

Problem (Implied)
"We could use either of two carbon papers; we don't know which is best."

Technical Questions/Tasks

Communication Purpose

The gas centrifuge promises to be an economical method for uranium isotope separation because of its low electrical requirements. Since several centrifuge designs operate successfully, it is necessary to determine which design offers the best solution in terms of reliability, ease of operation, and economy. This report compares the results of several leading centrifuge design contenders.

> *Background Problem*
> *We must separate uranium isotopes but this requires lots of electricity.*
>
> *Problem*
> *Several designes work; we don't know which works best.*
> *Technical Questions/Tasks*
> *Which works best by these criteria?*
> *Communication Purpose*
> *Implied but not precisely stated: "This report recommends the best choice."*

On February 13, 1980, the Naval Electronic Systems Command (NESC) rejected a Wylie Industries report which recommended modifying and updating MIL-STD-275C and MIL-STD-1495 to eliminate the requirement for solder plugs in plated-through holes in printed circuit boards. Wylie Industries had argued that because of modern materials and technology, solder plugs are no longer necessary to insure electrical continuity in printed circuit boards. NESC

> *Problem*
> *"We argued for a change, but they rejected our arguments."*

Fig. 3-3
Annotated Examples of Effectively Stated Purposes

rejected the recommendation for three reasons:

1) insufficient supporting data;
2) inadequate laboratory test;
3) NESC's body of data contra- dicting Wylie Industries data.

Technical Questions or Tasks
"Can you respond to these objections that they have raised? How?"

The purpose of this report is to respond to NESC's objections and to again recommend that elimination of the requirement for solder plugs will produce a less expensive, more reli- able circuit board.

Communication Purpose
"Please reconsider."

The acceptance test of the VPD pulser system was completed August 14, 1979. Preliminary analysis of the data indicates general compliance with the statement of work with exception of paragraphs 5.0, 6.0 & 9.0. Paragraphs 5.0 & 6.0 cover maximum output vol- tage & frequency content of output. Paragraph 9.0 deals with electromag- netic interference. It states, in part, "the pulser system shall be

Problem
We are generally but not completely in compliance

Technical Questions or Tasks
(Implied) "Are we close enough in compliance to recommended accept- ance?"

operated at full voltage without affecting its performance." The project office recommends conditional acceptance of the Pulser System, making note of the discrepancies.

Communication Purpose
"Please grant acceptance under these conditions."

Fig. 3-3
Annotated Examples of Effectively Stated Purposes

41

appear to be no valid reasons for explaining the problem and the assignment in detail, then simplify. But be sure you do not take short cuts when it is inappropriate to do so.

3.3.5 Location of the Purpose Statement

A purpose can be stated more than once in a report. Besides the first paragraph, you can clarify your purpose in two other places in most reports: in the subject line or title and in the concluding paragraph of the report.

Your reports will almost always have a heading that includes a subject line or a title. If so, you can use this subject line to clarify both the topic and the purpose of the report. It takes just a little attention to make the subject line serve both those functions.

Subject lines often state only the topic of a report, such as, "An Expanded Insulated Gate." This title gives you no clue to the purpose of the report, which in this case was a recommendation for a design change. In the case of the subject line, "Relative Humidity Failures," since the purpose of this report was to propose a solution to the failures, the subject line might have been, "Proposal for Eliminating Relative Humidity Failures." The word "proposal" tells you the communication purpose of the report and adds meaning to what otherwise is just a very general topic designation.

As you can see, it is easy to give both your topic and purpose in your subject line. We suggest you routinely try to state both in your subject line or title. Here are several additional examples of subject lines, revised to clarify both topic and purpose:

Original: Data Acquisition System
Revised: Data Acquisition System: Justification of Proposed Design

Original: Engine Accessory Mountings
Revised: Engine Accessory Mountings: Request for Information

Original: Speed Acceleration System
Revised: Proposal for Redesign of Speed Acceleration System

Original: Staffing in Emissions Laboratory
Revised: Request for Permission to Hire in Emissions Laboratory

With a little forethought, you can make your purpose evident to your readers even before the first paragraph.

Many reports can also have a concluding paragraph or sentence to refocus your readers' attention on the purpose. Your ending might be something like, "I would appreciate receiving your approval of my proposed design change" or "If you have any questions about my proposal to reduce the turnaround time in the X-ray department, please call me at 764-1427." In a longer report you might say, "Our study shows that the cutting tools used in the shop do not conform to OSHA safety standards. They must have safety guards and switches installed if we are not to risk being cited by the OSHA inspector. Cost would be approximately $630.00. I will go ahead with the modifications as soon as I have your approval."

The last paragraph of a report can serve as a restatement of the purpose and the important conclusions or recommendations the reader should really pay attention to. This gives you a graceful way to end the report and provides a little added insurance that you will indeed accomplish your purpose.

In summary, you must be able to understand and state the purpose for every report you write. If not, you risk writing reports which are ineffective in design, content, and method of presentation and which will not accomplish what you set out to do. The simple three-step method will assure that both you and your audiences will understand the purpose of your reports.

You must practice the method in every report you write until you become proficient at using it. Your reports will be easier to write and better received than if you neglect the distinction between your technical purpose and your communication purpose.

Exercises

A.   The three-step outline can be used to evaluate purpose statements. Using it as your guide, identify precisely how the following purpose statements are deficient and write a brief explanation in this format:

Statement (a, b, c, or d) ———
1. Problem:
2. Assignment:
3. Communication purpose:

a. One of our batch-trucks on the Ohio job recently encountered several mechanical difficulties. Mr. Holloway asked me to investigate the economics of repair compared to replacement and to send you my evaluation. The purpose of this report is to suggest that it would be unwise to repair the truck, because of its age, and to recommend that a new truck be purchased.

b. The Tri-County Hospital Council has started an annual survey of physical therapy departments. As a member of the council, Midland Hospital is participating in the survey. The questions cover organization, staffing, expenses, and activity. One question has caused particular problems for Midland's PT department. While PT has a simple reporting system which records type of treatments by patient by day, they currently have no data to determine direct patient contact time. The primary purpose of this project therefore is to conduct studies to obtain direct patient contact time.

c. The town council has authorized a new signal at the intersection of Front Street and Eighth Street. Before it can be installed, though, a traffic study has been requested. This study will determine the traffic patterns so that the signal can be pretimed for smooth operation.

d. General Motors has recently initiated a significant corporate tire-testing program. Improvement in the wet traction performance of the original-equipment tires is one of the goals of this program. General Motors' participation in wet traction studies has necessitated the development and construction of test surfaces which have controlled friction properties. This report concerns the design and building of one of these test surfaces.

B.   Write a subject line or title for each of the four reports excerpted in Exercise A and for the reports in Figure 3-3. Make both the topic and the purpose clear.

C.   Write a full three-step purpose statement for a report in your own area of specialization.

D.   Write a full three-step purpose statement for the report whose author you interviewed for Exercise A, Chapter 2.

4.1 Introduction

On a gray Monday morning, Len LaFara, manager of the sales division of a large company, looks dejectedly at the stacks of paper on his desk and slumps despairingly in his chair. How will he ever catch up?

His secretary, Ms. Ambrose, enters. "Here are the marketing reports you requested, Mr. LaFara." She drops them on his desk and leaves.

These reports will give Len the information necessary to make a vital decision before the afternoon meeting. He lifts a report and visibly shudders just at its weight. He reaches for his coffee.

Len slowly reads through the first five pages. His head starts to ache. "What a way to start the week," he mutters to himself. He turns a few more pages, then flips through several until he finds a heading, "Field Activities." The pain settles right behind his left eyeball. Even his master's degree in business management hasn't prepared him to cope with the statistical jargon, and who cares who was interviewed in Kankakee? What's the point of it all? His head throbs. He drops the report on the "to do later" pile and hopes there's nothing in it that would have helped him make a decision. He reaches for aspirin and the second report, in that order.

Len starts to read the second report and soon relaxes. He can even understand the first page. By the end of the page, he knows the company problem, the alternatives, and a possible solution. Len smiles contently, jots some ideas in the margin of the report, and puts it in his folder for that important conference in the afternoon. He will look good there. He notes the author of the report and says to himself, "Remember him when promotions come around."

Len's headache is gone. He buzzes Ms. Ambrose and asks her to remind Joe Minelli of their golf game at four.

As for that first report, it sank deeper and deeper in the "to do later" pile with each passing day, until one day Ms. Ambrose gathered it up and took it to the files in the documentation room down the corridor. Len LaFara never did find out who wrote it.

How do you write a report that will be read? This is more difficult than it sounds. If you haven't put on your writer's hat, you'll tend to do everything backwards. You'll start at the beginning of your technical activities, launch into the particulars, and present a step-by-step account. You'll describe what you did this way because that is easiest and seems the natural thing to do. Forced to wade through a bog of details, your readers will become impatient and often miss your most important points.

To write a report that will be read, you must put on your writer's hat and write to meet your readers' needs. You write a report so your readers can best understand it and use it most efficiently. Your first task as a writer is to design an effective basic structure. Accordingly, the objectives of this chapter are to enable you to:
1. Understand that your basic report structure has two segments.
2. Arrange your information so it moves from general to particular between these two segments.
3. Write the opening segment of a report.

As we explain how to accomplish these objectives, we will draw upon the principles established in previous chapters. Your ability to analyze the audiences for a report and to write a purpose statement will help you to design the basic structure of a report.

4.2 The Two Basic Segments of a Report

You seldom find anything that consists of just one structure—a single collection or sum of parts. For example, an automobile consists of the power plant, which in turn consists of the pistons, block, etc.; a body, which consists of the doors, hood, and so on; a chassis and suspension system; and other substructures. The automobile is not the sum of the instrument panel, the springs, the doors, the cylinders; instead it is a combination of various substructures. A football game consists of two halves, each with its own sequence of action. Many a coach has rued the afternoon that a football game did not go from beginning to end without a half time. A hospital consists of admissions, offices, diagnostic stations, laboratories, patient rooms, etc. As most of us know, each subdivision of a hospital has its own procedures and waiting times. Thus, while an automobile, a football game, and a hospital may seem at first to be single things, we invariably discuss them in terms of one or more of their substructures. A report also should be thought of as a collection of substructures rather than as a single series of words, sentences, and paragraphs from beginning to end.

We can think of a report as having two basic substructures, two basic segments: the opening segment and the discussion segment (Table 4-1).

You should be able to see the break between the opening segment and the discussion segment in any effectively designed report. In fact, you should be able to read far enough to get the main point, take scissors, and cut the report there. You will have an opening segment that by itself meets all the information needs of some report readers. If you can apply this scissors test to your report, you have mastered the first step in basic report design: thinking in terms of two basic segments.

Suppose you are a laboratory technician in a soils laboratory, working for Mr. Robert Jones, a consulting engineer and director of the laboratory. He has been asked if the soil at a construction site will support the proposed facilities. You go to the site, perform the standard soil-density tests, and examine the construction plans. You then compare your results with handbook standards, check your conclusions with your supervisor, and write a memo to Mr. Jones, who will attach it to his letter of response to the construction company, American Manufacturing. (In many situations you will actually write those letters yourself, but they will be sent out over the signatures of the managers and directors.)

The memo you write might be like that shown in Fig. 4-1. Notice that this memorandum has two basic segments. Apply the scissors test to this report and examine the opening segment:

On September 23, 1975, American Manufacturing Co. requested field-density testing at their proposed Delhi Road plant site. They expressed concern about the possibility of soil settlement due to the vibration of high-speed punch presses. I performed standard density tests, and find no cause for concern.

The tests indicate that the soil of the Delhi Road site is naturally in a highly densified condition. Virtually no settlement will occur at this site under such a loading. This site is an excellent location for the placement of vibratory machinery.

As you can see, this opening segment can stand by itself. It is intelligible to any reader and does not force the reader to go into the discussion segment in order to get the message. It effectively meets the needs of management at American Manufacturing, who only want the bottom line: "This site is an excellent location for the placement of vibratory machinery."

If you had not thought in terms of two basic segments, you might have written the report as follows:

On September 23,1975, American Manufacturing Co. requested field-density testing at their proposed Delhi Road plant site. They expressed concern about the possibility of soil settlement due to the vibration of high-speed punch presses.

I performed tests on both a one-foot layer of dense, brown, clayey sand and a dense, uniform, fine sand found beneath the surface layer at the Delhi Road site. The in-situ density was determined by the "sand cone" and "drive sampler" methods. These were the standard ASTM tests D1556 and D2937 respectively.

For the fine sand, the following results were computed:

Dry density: 102#/ft.^{3}
In-situ void ratio: 0.61

Table 4-1
Two Basic Segments of a Report and Their Functions

Segment	Parts of the Report	Functions
Opening	Preliminaries and Summary	Provides an instrumentally useful overview of the problem and solution
Discussion	Discussion	Provides detailed discussion of the problem,
	Appendices or Attachments	technical investigation, and solution

SOILS LABORATORY

Dexter, Michigan

October 1, 1979

To: Mr. Robert Jones, Director

From: John O'Brien, Soils Analysis

Subject: Field Density Tests at Delhi Road Site for
          American Manufacturing:

On September 23, 1979, American Manufacturing Co. requested
field density testing at their proposed Delhi Road plant
site. They expressed concern about the possibility of soil
settlement due to the vibration of high-speed punch presses.
I performed standard density tests and find no cause for
concern.

The tests indicate that the soil of the Delhi Road site is
naturally in a highly densified condition. Virtually no
settlement will occur at this site under such a loading.
This site is an excellent location for the placement of
vibratory machinery.

I performed density tests on both a one-foot layer of dense, brown
clayey sand and a dense, uniform, fine sand found beneath
the surface layer at the Delhi Road site. Both layers had
relative densities of close to 100%. Since construction
standards consider a relative density of 70% adequate for
vibratory loads, very little settlement would occur at the
Delhi Road site.

The in-situ density was determined by the "sand cone" and
"drive sampler" methods. These were the standard ASTM
tests D1556 and D2937 respectively. For the fine sand, the
following results were computed:

$$\text{Dry density: } 102\#/\text{ft.}^3$$
$$\text{In-situ void ratio: } 0.61$$

The results of the clayey sand were not considered significant since excavation plans call for this top layer to be
removed before construction.

Fig. 4-1
Specimen Laboratory Memorandum on Soils Test

The Technician as Writer

The results of the clayey sand were not considered significant, since excavation plans call for this top layer to be removed before construction. Both layers had relative densities of close to 100%. Since construction standards consider a relative density of 70% adequate for vibratory loads, very little settlement would occur at the Delhi Road site.

The tests indicate that the soil of the Delhi Road site is naturally in a highly densified condition. This site is an excellent location for the placement of vibratory machinery. Virtually no settlement will occur at this site under such a loading.

In this particular-to-general order the report fails the scissors test. It can't be cut in two; it forces all readers to read to the last line to get the main point. The report treats all readers equally, rather than addressing their individual needs.

4.2.1 Readers' Needs in Each Segment

Each segment of your report addresses different types of readers or—to put it differently—addresses different reader needs. Many readers need to know only the main point of a report, without the particulars. The opening segment addresses these audiences. Other readers, however, need to know something about the particulars of your technical or professional activity. The discussion segment is for these readers.

Previous chapters have introduced you to the egocentric organization chart as a means of identifying the various audiences any report can have. With this background, you ought to be able to identify two possible groups of readers for the soils-test report, including:
Laboratory supervisor, the writer's boss
Laboratory director, Mr. Jones
Vice-president for plant facilities, American
   Manufacturing
Director of plant construction, Delhi Road plant
Construction engineer, Delhi Road plant
Construction supervisor, Delhi Road plant

Classify the needs of those people according to whether they will read only the opening segment or must also read the discussion segment. A possible classification might be:
Opening Segment Audiences
   Laboratory director, Mr. Jones
   Vice-president for plant facilities, American
      Manufacturing
   Director of plant construction, Delhi Road
      plant
Discussion Segment Audiences
   Laboratory supervisor
   Construction engineer, Delhi Road Plant
   Construction supervisor, Delhi Road Plant

The opening segment audiences have no need for the particulars. They are not familiar with "sand cone" and "drive sampler" methods, nor would they know if "in-situ void ratio: 0.61" was good or bad. These persons depend on others to know such particulars. They need only the answer to the question, "Can we build the plant there?"

The discussion segment audiences have some need to know particulars. The laboratory supervisor needs to verify your conclusions. After all, his or her judgment is on the line, too. The construction engineer and construction supervisor have some interest in the particulars because they want to know the characteristics of the soil at the construction site. They probably have no concern for what is going into the plant.

Can you identify the predominate types of audience for the opening segment and for the discussion segment? The audiences for the opening segment are often in decision-making and other managerial roles. These audiences need to keep their fingers on the pulse of the organization. They must have an overview of many activities in order to be able to interact with other units. The audiences for the discussion segment often are in staff positions and hands-on operations. They work for others, performing specific tasks. These persons need particulars in order to perform their duties within their offices and units.

4.2.2 Variations in the Readers of Each Segment

It is clear that the two segments of a report address different audiences. However, do not assume that the opening segment always addresses primary audiences and the discussion segment always addresses secondary audiences. Your report purpose determines your primary and secondary audiences. Your primary audiences may be either management or line personnel. Thus the opening segment can address either the primary or the secondary audiences, and the discussion segment can address either type of audience.

Two specimen reports illustrate these alternatives. The "Specimen Laboratory Memorandum on Soils Test" has the following design:
Opening Segment: Aimed at primary audience of decision makers.
Discussion Segment: Aimed at various secondary audiences.

The "Specimen Memorandum of Instructions by an Electronics Technician" (Fig.4-2), however, has this design:
Opening Segment: Aimed at secondary audience of supervisors.
Discussion Segment: Aimed at primary audience of parts assemblers.

The purpose of "Specimen Memorandum of Instruction by an Electronics Technician" was to have production-line assemblers change their method of forming leads on plastic triacs. The primary readers are addressed by the discussion segment, not the opening segment, because they need particulars in order to follow the procedure explained. The opening segment of this report addresses secondary

Table 4-2
Alternative Audiences for Report Segments

Basic Structure for Report Aimed at Management

Segment	Units of Report	Audiences
Opening	Heading, Purpose, Conclusions, Recommendations	Primary audiences in management who need a useful overview of problem and solution; immediate audience of transmitters
Discussion	Particulars	Secondary readers, persons and units to be affected by implementation of conclusions and recommendations, as well as some management personnel
	Appendices Attachments	

Basic Structure for Report Aimed Primarily at Persons Without Administrative Responsibility

Segment	Units of Report	Audiences
Opening	Heading, Purpose, Conclusions, Recommendations	Secondary audiences in management who need only an overview of the information; immediate audiences of transmitters
Discussion	Particulars Appendices Attachments	Primary audiences of persons and units who will act as a result of the report

Dexter Electronics

INTEROFFICE MEMO   January 23, 1980

TO: F. Frauhammer, Production Manager
           W. MacMartin, Production Supervisor
           Production Foremen

FROM: D. Harger, Shop Technician

SUBJECT: Forming Leads for Plastic Triac Packages

REF: D E Product Notebook (B-39, D-3T, D-31T, D-31RT)

Our in-line plastic triac packages occasionally are returned because
of internal damages. I have determined that this results from
incorrect forming of the leads in the shop. The following procedure
therefore should be used to form the leads.

Problem. The plastic package can sustain internal damage when the
lead is incorrectly formed. If the lead is not restrained between
the bending point and the case, when being formed the lead can flex
at the case and cause internal damage (see Figure a).

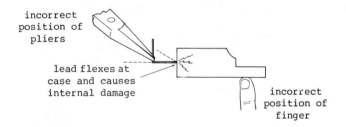

         Figure a. Unrestrained lead can flex at the case
                    while being formed.

Solution. The lead should be formed by restraining it between the bending point and the case when it is formed. This requires correct positioning of pliers and finger and correct tolerances when bending.

1. To restrain the lead between the bending point and the plastic case, position the pliers at the bending point and on the inside. Position the finger on the outside. (See Figure b).

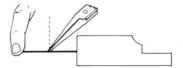

Figure b. Correct positioning of finger and pliers to form lead.

This will prevent relative movement between the lead and the case.

2. Use correct tolerances to form a sharp right-angle bend. The bend must be at least 1/8 inch from the plastic base and at least 1/16 inch from the end of the lead (Figure c).

Figure c. Correct tolerance for bending lead.

3. With the finger, bend the lead in a plane perpendicular to the plane of the base of the plastic package to form a right angle (Figure d).

Figure d. A right-angle bend should be formed.

4. Avoid repeated bending of leads.

This procedure should be posted in the shop and all personnel
reminded of the correct method of forming leads for the plastic
triac packages.

D. Harger

D. Harger

Shop Technician

                        Approved by:

                        F. Frauhammer

                        W. Mac Martin

Fig. 4-2
Specimen Memorandum of Instructions by an
Electronics Technician

audiences, mostly supervisors. The supervisors need only to know production techniques are being changed and the purpose of those changes; they do not need to learn the particulars of the procedures.

These two alternative patterns for basic report structure are presented in Table 4-2. The first alternative is a report aimed primarily at management. The opening segment of such a report addresses a primary audience. Management needs only general information in order to act and make decisions. The discussion segment addresses various secondary audiences, who need to implement the decision, will be affected by the decision, or need to analyze the particulars to document the conclusions and recommendations in the opening segment.

The second alternative is a report aimed primarily at persons without administrative responsibility. The opening segment of such a report addresses secondary audiences, usually management, who need only general information in order to keep abreast of what's happening. The discussion segment addresses the primary audiences, who must act on the basis of the particulars presented in the report.

For both types of reports, immediate audiences need only to scan the opening segment in order to get the report or the information it contains into the right hands.

4.3 General-to-Particular Organization

Our discussion of the audiences for the opening segment and discussion segment indicates another feature of basic report structure. The two segments of a report contain different types of information. The opening segment contains *general* organizational and technical information. The discussion segment contains *particular* technical information. This distinction is illustrated in Table 4-3.

Table 4-3
Types of Information Contained in Each Report Segment

Segment	Type of Information
Opening	General information on the organizational problem, technical investigation, conclusions, and recommendations
Discussion	Particular information on the technical problem, investigation, results, and analysis

4.3.1 Thinking from General to Particular

It isn't easy to move from general to particular. In fact, this requires you to reverse your thinking, because you do most of your technical activities by working in just the opposite order: particular to general. For example, if you are preparing an inventory report, you itemize the stock on hand and then add up the figures to arrive at your conclusion. To order supplies for the pharmacy department, you survey the items in stock, project the demand for the coming month, and decide how much of each item to request on the purchase order. When you write a report, if you are thinking of your work but not thinking of your audiences, you might find it easiest to go from particular to general, rather than from general to particular. But if you arrange the report in the same order in which you performed your technical activities, you will force all of your readers to wait until the end of the report to find out the most important information. If, however, you arrange the report in the reverse order, you will give your readers the important information at the beginning of the report.

Regardless of its purpose, almost every report needs to have a generalization first. A manager's first reaction to a report is, "What is this? Do I read it, route it, or skip it?" The generalizations tell him or her immediately. A supervisor first wants to know, "Who should get this report?" and "What will happen when I send the report?" The generalizations, not the particulars, answer the supervisor's questions. Even readers who need the particulars need the generalizations first, to enable them to read efficiently and selectively. A state inspector visiting a local agency, for example, appreciates being told first that she will meet with the department heads and then interview the case workers, and also that the writer has made arrangements for her visit. The inspector does not want to be told to take a limousine to the Holiday Inn, where arrangements have been made for her to stay. A respiratory-therapy technician receiving instructions on how to communicate with physicians and nurses first needs to know the four types of communication problems and purposes and how these differ from each other. Only after this has been done should the technician be introduced to the particulars of the "Adverse Reaction" memorandum or the "Discontinuation of Order" memorandum.

So you work from particular to general, but you write from general to particular. Your work may follow this particular-to-general pattern:

This patient complained of tremors, numbness in the hands, and a "funny feeling in the chest."

The pulse at start was 92.

After five minutes of IPPB, the pulse increased to 136.

Treatment was stopped and RN informed.

Should this patient continue treatments?

A progress note to the doctor, however, should go from general to particular:

Monday 10:35 a.m.
Doctor:
Should this patient continue treatment? During treatment, his pulse increased and he experienced discomfort. I stopped treatment and informed the nurse.

At start of IPPB, pulse was 92. After five minutes, the pulse increased to 136. Patient complained of feeling tremors, numbness in the hands, and a "funny feeling in the chest."

Elaine Black, CRIT
Respiratory Therapy
Department

The doctor should be told immediately that the report requires a decision. To do that, the report moves from the important general information to the particular evidence that something may be seriously wrong with the patient.

4.3.2 Writing from General to Particular

To reverse your thinking from particular to general to general to particular requires you to turn from your work and face your readers.

In order to write general to particular, of course, you must first be able to distinguish between the particular and the general. You also must be able to arrange information according to levels of generality.

The distinction is between what can be observed directly (particulars) and what cannot be observed directly (generalities). Perhaps, for example, the County Parks Commission needs population-density statistics, and you are asked to report the population density in Webster Township. You tabulate the exact number of persons living in each of the thirty square miles of the township. This tabulation represents the census takers' direct observations of the persons living in Webster

Township. This is particular information. When you average statistics, however, you arrive at general information. When you say that the population density in Webster Township is 32.7 persons per square mile, you are presenting information that cannot be directly observed.

The most particular information takes the form of direct empirical data and observations: measurements, counts, direct impressions of the senses, details, and calculations. Here are examples:

The planned park is six miles from Webster.

All fractures occured on test units run less than 5,000 miles.

The dry density was 102#/ft.³

The patient received 30 mg. of Bronkosol at 6:00 a.m. and at 6:00 p.m.

The drive-in window line, between 10:35 and 11:00 a.m. on Saturday, May 2, extended onto Packard Street up to six cars.

Generalizations come from averaging, comparing, interpreting, inferring, evaluating, judging, and concluding. Notice that in all of these, one or more human beings mentally process the particulars. Here are examples of generalizations:

The planned park will be accessible to Webster inhabitants.

The planned park is close to Webster.

The planned park is legal; that is, it is more than five miles from a section with a population density of more than 100 persons per square mile.

All fractures occurred in low-mileage test units.

Fractures will occur before 5,000 miles or not at all.

All units with mileage under 5,000 miles should be tested.

The relative density was close to 100%.

The soil is very dense.

Virtually no settlement of the soil will occur.

The patient received Bronkosol as directed.

The patient received normal Bronkosol therapy.

The line extended several cars onto Packard for about a half hour.

You must separate the general from the particular, and put the general first. If your results show that the planned park is six miles from Webster and your conclusion is that the planned park is accessible to Webster inhabitants, you state your conclusions before going on to discuss the particulars:

The Webster Town Council asked the County Parks Commission if the planned park would serve the citizens of Webster effectively. I checked our guidelines and the legal requirements. Here is the information for your May 3rd meeting with council members.

The planned park is accessible to Webster inhabitants according to our standard guidelines for park planning. It will service the town of Webster according to state tax laws as well. The relative accessibility of the park to the residents of Onsted in York Township may be the real issue.

Criteria for accessibility. Commission guidelines . . .

To be able to write this way, you must be able to realize that "The planned park is accessible to Webster inhabitants" is a generalization from the particular fact that the planned park is six miles from Webster. You select general statements that express your interpretations of the specific information, rather than simply stating the specific information itself.

After you separate the particular from the general, you need to distinguish between levels of generality. The statement "The population density poses no immediate pollution threat to Cross Lake" exhibits a significant degree of interpretation and is therefore quite removed from the particulars. The statement "The present population density is highly desirable" is a generalized value judgment. Notice, however, that these two statements differ in their degrees of generality. The second statement is more general than the first. You need to arrange statements according to how general they are.

We distinguish among three types of generality: results, conclusions, and recommendations, requiring different degrees of interpretation by you, the technician. Some statements require you to interpret particulars objectively. Other statements, however, require you to exercise considerable interpretive or critical judgment.

A *result* requires routine mental processing of data. The statement that the soil density is 102#/ft.³ is a presentation of measured data. When you state that the relative density is close to 100 percent, you are making a low-level generalization, involving the objective combining and averaging of data. A low-level generalization is a statement of information that cannot be observed directly, but at the same time requires little human judgment or evaluation. Low-level generalizations are objective interpretations of empirical data.

A *conclusion* requires subjective interpretation of results. When you state that the soil is highly densified, for example, you are interpreting a result, you are drawing a

conclusion from it. In this case, you have derived a subjective concept, high density, from the results. When you say that virtually no settlement will occur, you are making a deduction. Such processes as *concluding, deducing, inducing, inferring,* and *judging* result in high-level generalizations, because you view the empirical data and results in a context of criteria and standards.

A *recommendation* requires evaluation of a conclusion in an organizational context. For example, when you assert that a site is an excellent location for the placement of vibratory machinery (because virtually no settlement will occur), you are evaluating a conclusion in a context requiring decision and action.

Table 4-4 presents examples of each level of generalization. It is not easy to arrange information according to these types, however, because the boundaries are not sharp. For example, the following three statements illustrate how types of generality actually are interpretations along a continuous spectrum. Statement 2 does not seem to be much more general than Statement 1, nor does Statement 3 seem much more general than Statement 2:

1. The soil has a relative density close to 100%.
2. The soil is in a highly densified condition.
3. Virtually no settlement will occur at this site.

A close examination, however, indicates that Statement 1 is closer to particulars than Statement 2. Statement 2, while including the value judgment that "close to 100%" is "highly densified," in turn lacks the deductive generalization contained in Statement 3. This deductive content is revealed when Statement 3 is expressed in another context as: "Since a relative density of 70 percent is considered adequate for vibratory loads, very little settlement would occur at the Delhi Road site."

As a report writer, you should learn to arrange information according to levels of generality by signalling what interpretation or evaluation you have applied to the results. Such terms and phrases as "highly" or "since a relative density of 70 percent is considered adequate for vibratory loads" tell the readers what guidelines have been used.

You also must recognize higher levels of generality that may be implicit in your results. A conclusion often contains an implied recommendation. For example, the statement, "We consider this site an excellent location for the placement of vibratory machinery," implicitly recommends the site. In making this statement, the soils laboratory was well aware that its reputation was on the line. A laboratory technician testing food samples also should be aware of what he or she is implicitly recommending when certifying food specimens according to the health code.

When a result suggests a valid conclusion, the report writer should state the conclusion explicitly. For example, these results signal a conclusion that wasn't stated:

The second cash register increased customer check-out from 12:00 to 1:00 p.m. by 32%.

The average wait was four minutes, and the average line length was eight persons.

The implicit conclusion was that two cash registers do not meet the objectives for customer check-out. The report writer unfortunately failed to generalize to that extent, even though the organizational problem required the conclusion to be made explicit and the store manager obviously was interested in the conclusions rather than the details.

Report writers often feel uncomfortable about going beyond the objective facts. When your audience analysis and purpose statement tell you what your audience needs to know, you should draw the appropriate conclusions from the data. A positive evaluation of your job performance by your supervisor even may depend on your ability to generalize appropriately, that is, on your ability to generalize to the degree warranted by the situation. To write general to particular, you must generalize.

Table 4-4
Levels from Particular to General for Report Writers

Level	Examples
Empirical Observations (direct measurement, data, sensory impressions, details)	The layer of fine sand had a density of 102#/ft.³ The add-on heat shield reduced the engine housing inside surface temperature from 475°F to 292°F. Men's Clothing had 62 returns in January.
Results (objective interpretations of empirical data)	Both layers had relative densities of close to 100%. The carburetor changes account for a 70°F increase in inside surface temperature. Men's Clothing had the highest proportion of returns of any department in January.
Conclusions (application of values, criteria, and standards to empirical observations and results)	Virtually no settlement will occur at this site under such a loading. The G-22 engine housing has design deficiencies. The rate of returns in Men's Clothing exceeds company standards.
Recommendations (evaluation of conclusions in organizational contexts)	We consider this site an excellent location for the placement of vibratory machinery. The add-on heat shield should be released for production. The cause of excessive rate of returns in Men's Clothing should be investigated.

The Technician as Writer

This discussion helps you solve a problem that may not have been immediately apparent in the explanation of basic report structure. The basic report structure consists of two segments, as illustrated in Table 4-3. The opening segment gives general information on the organizational problem, technical investigation, conclusions, and recommendations. The discussion segment, on the other hand, provides particular information on the problem, investigation, results, and analysis. The opening segment should consist primarily of high-level generalizations—conclusions and recommendations. The discussion segment should consist primarily of low-level generalizations—analysis of results. Empirical observations go in the appendices, the attachments, and the file. Formulate the appropriate conclusions and recommendations, and insert them in the opening segment. Put your results into the discussion, analyzing them there.

4.4 Designing the Opening Segment

The opening segment is the most important part of your report to many readers. You must learn how to write an effective opening segment if you are going to write good technical reports.

An opening segment has three units, the heading, the purpose, and the conclusions and recommendations. These sometimes are explicitly identified on the organization's report form; at other times they are not, as in many letters and memoranda. In either case, you should consider each unit separately so that you make sure you include the appropriate information. You will find that considering each unit separately makes report writing easier. These units of the opening segment have different functions.

4.4.1 The Heading

The heading attracts the attention of the appropriate readers in your organization, so that they pay attention to your report. The heading introduces the report to the organization. In order to do its job, the heading must contain certain information:
1. The subject of the report.
2. The audiences for the report.
3. The source of the report.
4. Reference information about the report.

When this information is present, a reader can determine what the report means to him or her and what it does for the organization. Some readers need to read only the heading of a report.

The subject is often set off as a title. The memorandum format usually is:

SUBJECT: Field Density Tests at Delhi Road
Site for American Manufacturing

This same subject line could be presented by itself, either flush with the left margin or centered:

FIELD DENSITY TESTS AT DELHI ROAD SITE
FOR AMERICAN MANUFACTURING

The subject line or title of a report should be as precise as possible, clearly stating the topic and purpose of the report. All nouns and verbs should be important, specific, and concrete. General terms such as "study," "analysis," and "report" should be omitted. Examples of effective subject lines are shown in Fig. 4-4 and 4-5.

The audiences of the report should be identified as completely as possible. Reports that are issued with insufficient audience

```
┌─────────────────────────────────────────────────────────────────────────┐
│                PROVING GROUND TRUCK DEPARTMENT REPORT                     │
├──────────────────────────────────────────────────────┬──────────────────┤
│ SUBJECT                                                │ PAGE             │
│                                                        │                  │
├───────────────────────────┬──────────┬────────────────┼──────────────────┤
│ REPORTED BY               │ DATE     │ TEST REQUEST   │ FILE CODE        │
│                           │          │                │                  │
├───────────────────────────┴──────────┴────────────────┴──────────────────┤
│ PURPOSE:                                                                  │
│                                                                           │
│                                                                           │
│                                                                           │
│                                                                           │
├───────────────────────────────────────────────────────────────────────────┤
│ CONCLUSIONS:                                                              │
│                                                                           │
│                                                                           │
│                                                                           │
│                                                                           │
│                                                                           │
│                                                                           │
│                                                                           │
│                                                                           │
│                                                                           │
│ RECOMMENDATIONS:                                                          │
│                                                                           │
│                                                                           │
│                                                                           │
│                                                                           │
├───────────────────────────────────────────────────────────────────────────┤
│ CC                                                                        │
│                                                                           │
└───────────────────────────────────────────────────────────────────────────┘
```

Fig. 4-3
A Standard Report Format Illustrating the Three
Units of the Opening Segment: the Heading, the
Purpose, and the Conclusions and
Recommendations

information can fail to alert the readers to the uses and importance of the report. Reports that address only immediate audiences rather than the primary audiences can be ineffective because it is possible that neither will recognize who should pay attention to the report.

Unless company formats and protocol prevent you from doing so, you should address the report to its primary audiences. This is especially important for internal reports. When possible, address the report to several audiences, listing the most important audience first. Important secondary audiences can be listed after the primary audiences. Other secondary audiences can be identified on copy lists or by other formatting devices.

You must identify the audiences by their roles and offices, as well as by their names. The usual sequence is author, role, unit. Your readers need to know the roles in order to grasp the purpose of your report quickly. Role identification is particularly important for reports copies of which go outside the organization or to distant offices within the organization. Examples of effective audience address are shown in Fig. 4-4 and 4-5. Readers need to know the source of the report, as well as the audiences, in order to grasp the purpose of the report. This means stating the writer's name and role, as well as which organizational unit he or she belongs to. The actual writer should be identified if possible, but some organizational protocol requires all reports to be issued under the name of the unit supervisor. In these instances the real author sometimes is identified by a formatting device, being listed second or identified by a phrase such as "per John Doe" at the end of the report. When the issuing unit is not already identified on the stationery letterhead or memorandum report form, it also should be stated in the report.

Identifying the writer by role and unit is important, because some readers, especially distant ones, recognize roles, but do not recognize names. Also, people's roles change as the writer gets promoted or moves to another office. Examples of effective source identification are shown in Fig. 4-4 and 4-5.

You provide reference information to place the report in time. The date of issuance by itself is not sufficient. Previous memos should be identified by date or special reference numbers when the audiences need to know these to comprehend the purpose of your report completely. File numbers, key words, and retrieval codes enable the report to be accessed in the future. Such reference information identifies the report as a response to past reports, as functioning in the present, and as useful in the future. Examples of effective reference information are shown in Figs. 4-4 and 4-5.

4.4.2 The Purpose Statement

The purpose statement, sometimes called the foreword, is the first paragraph in your report. As we saw in Chapter 3, this statement enables the readers to understand the organizational context of the report and its communication purpose. Those readers need to know the organization's problem and to anticipate the uses of the report. The purpose statement immediately orients the reader to the type of response he or she must make.

The purpose statement of the report introduces the conclusions and recommendations paragraph, which follows immediately. Thus, the first two paragraphs of a report provide a useful summary of the problem and the form A-1048

INLAND WATERS

Juniper Bay Test and Maintenance Facility

Source of report issuing unit—but not the writer

primary audience

TO	B. Skinlea
TITLE	Vice President
DEPT.	Product Development

FROM *E S Valery*

approval for issuance

secondary audience

COPIES	Dept. 32 Marine Engineering
	Fuel Economy Task Force
	Consumer Products

TITLE	Head
DEPT.	Power Plant Testing
REF.	Test Request 7932

reference

SUBJECT   Alternative Blade Configurations to Improve Fuel Economy

subject states topic and strongly applies purpose

Foreword

Inboard Panther fuel economy improvement is required to meet
Inland Waters's 1982 goals. The Fuel Economy Task Force
requested vessel tests be conducted with alternative propeller
configurations to determine fuel economy improvement. Marine
Engineering therefore issued Test Request 7932 specifying the
propellers to be tested. This report presents the results of
the tests and recommendations to improve fuel economy.

problem

technical investigation and assignment

purpose of report

Summary   *provides overview / documents the recommendation*

Back-to-back tests on fuel economy and noise levels were
conducted on the two-blade and the four-blade alternatives
proposed to replace the standard three-blade propeller. Test results
showed improved fuel economy for the two-blade installed on the

Fig. 4-4
Opening Segment from a Test Report, with
Necessary Information Identified

The Technician as Writer

260 hp power plant. Results also indicated no fuel economy improvement for the four-blade on the 375 hp plant. Because of improved fuel
economy as well as acceptable noise levels, the two-blade
propeller should be released on the 260 hp plant.

Introduction

Rising fuel costs and consequent potential cust mer co rn
necessitate evaluation of potential fu l saving d
the 1982 inboard mari e market. Te
specified lt at l
eff dered for the
guration can be improved
y.

3. Test Request 7932 is terminated with issuance of this
report.

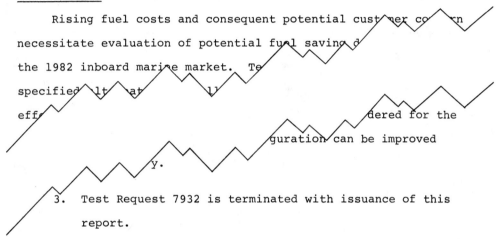

per. Alvin Karras

} *Writer of report*

Fig. 4-4
Opening Segment from a Test Report, with
Necessary Information Identified

ANN ARBOR GENERAL HOSPITAL

Inter-Office Correspondence

primary audience }

UNIT	LOCATION	DATE
Administration	A6018	May 19, 1978

NAME
S. A. Kingsfield, Administrative Manager

writer }

FROM	DEPT. & LOCATION
J. C. Ruffing, Systems Analyst	Operations A5006

} issuing unit

SUBJECT

Hospital Printing Capacity Recommendation: Xerox 9200 copier. *{ subject states substance and purpose*

Purpose. The addition of the new outpatient wing to the hospital *} organizational problem* has resulted in an increased amount of paperwork associated with processing these additional patients. This can be met by adding *} technical questions* an operator during the second shift or by upgrading the copying system. A cost analysis indicates that a new copier should be *} purpose of report* leased.

Recommendation. To meet a projected printing requirement of 900,000 copies per year, a Xerox 9200 copier should be leased. *} selects the most important points* This copier has an automatic document feed and an automatic collator. The total monthly cost will be $1,925.00/month. The existing printing system with overtime added would be about $2,465.00/month.

Benefits of the Xerox 9200 Copier. The Xerox 9200, by virtue of its automatic document feed and automatic collating capabilities, will provide the necessary production to meet these increased printing demands. The automation feature of the Xerox 9200 in-

Fig. 4-5
Opening Segment of a Proposal, with Necessary
Information Identified

solution for every reader. Many readers will need to read no further.

Appropriate information for the purpose statement is:
1. Organizational problem—the conflict at issue and its organizational context.

2. Technical investigation—the technical problem and the specific technical questions or task it required, which can be stated as the writer's specific assignment.

3. The communication purpose of the report—how the report should be used or what the impact of the report will be.

For most memoranda and reports of less than four pages, the purpose statement must be very concise. Usually the purpose statement will be a short paragraph of perhaps three or four sentences; in a one-page report, it can be perhaps just two sentences.

In some reports, you need to explain the organizational issue in detail or to clarify the technical problems and investigation. If this is necessary, do it at the beginning of the discussion segment rather than in the purpose statement.

As we saw in Chapter 3, you will often present the purpose statement in the sequence: 1. organizational problem, 2. technical investigation, 3. purpose of the report. This sequence is not mandatory, however. Many effective purpose statements begin with a statement of the purpose of the report, for example. In most reports, the subject line also can convey some of the purpose statement information; in some very brief memos, the subject line serves as the only purpose statement. Examples of effective purpose statements are shown in Figs. 4-4 and 4-5.

4.4.3 Conclusions and Recommendations.

This paragraph or, in reports of several pages or more, sequence of short paragraphs presents a condensed statement of the most important information. It presents the solution to the organizational problem identified in the purpose statement. The conclusions sum up your technical investigation; the recommendations evaluate the conclusions in the organizational context.

The conclusions and recommendations unit, sometimes labelled the summary, enables some audiences to act without reading further in the report and provides a useful overview for those readers who need to read the discussion. Because the primary purpose of the conclusions and recommendations is to enable some audiences to act, however, the information must be selected very carefully. The conclusions and recommendations unit focuses on the most important information rather than merely summarizing the discussion. In that sense, the word "summary" is misleading. Better would be "Summary of Important Conclusions and Recommendations."

Do not introduce particulars of the results and investigation into this unit. Select only the details that are absolutely necessary for your audience to understand your conclusions and recommendations. Many writers overwrite this unit, because they feel compelled to start explaining in the opening segment. They forget that the discussion segment will analyze and explain the conclusions and recommendations. Readers who need detailed explanation will refer to the discussion; readers who don't need these details should be presented directly with the conclusions and recommendations.

The conclusions and recommendations unit can repeat certain important organizational information almost word for word from the discussion. Important recommendations, costs, benefits, and subsequent action are sometimes directly repeated, for example. In some reports, costs and benefits are important enough to be set off in a paragraph following the recommendations.

Conclusions and recommendations usually are presented in that order, although some reports present recommendations first and then support them by the conclusions.

A typical outline for this unit is:
Methodology and Basic Results (if necessary)
Conclusions
Recommendations
Costs and Benefits

The methodology or basic results should be a summary of information, presented only to clarify the conclusions. Examples of effective conclusions and recommendations are shown in Fig. 4-4 and 4-5.

The explanation in this chapter applies to all the reports you write, regardless of format. The basic structure is sometimes explicitly formatted by the organization, as in Fig. 4-6. At other times, you must format the basic structure yourself by introducing the appropriate headings for the opening segment paragraphs, as in Figs. 4-4 and 4-5. In letters you may not format your basic structure at all, but you should introduce headings even into letters whenever that seems appropriate.

It is important for you to understand the basic structure of any good technical report, no matter what its particular format may be. When you are given an explicit memorandum format to work with, you still must design the basic structure.

PROVING GROUND TRUCK DEPARTMENT REPORT			
SUBJECT			
REPORTED BY	DATE	TEST REQUEST	FILE CODE
PURPOSE:			
CONCLUSIONS:			
RECOMMENDATIONS:			
CC			

LABORATORY TEST	
SUBJECT	PAGE
PARTS TESTED	
TEST PROCEDURE	
TEST RESULTS	
DISCUSSION	

Fig. 4-6
Standard Memorandum Forms (Extra Pages Added as Necessary)

7/6/78
Judith Nestel

ATTRITION STUDY

This study presents statistics on the attrition of full-time students at Toledo
County Community College from 1975 to 1977. The number who matriculated as well
as the number who graduate after more than four semesters also are presented.
This study also presents statistics on the attrition of part-time students that
the first study did not include.

The permanent records of full-time students who matriculated in associate degree
programs for the fall semesters of 1975, 1976, and 1977 were reviewed for purposes
of this study.

Listed below is the number of students selected at random from the total population.

MATRICULATION DATE	SAMPLE SIZE	POPULATION	PERCENTAGE
Fall 1975	362	923	39
Fall 1976	291	683	43
Fall 1977	310	731	42

The following table lists the number of student withdrawals by semester, percentage,
and accumulative percentage.

Table of Withdrawals Prior
to or During the Listed Semesters

	Fall 1975 # of Students	%	A-%	Fall 1976 # of Students	%	A-%	Fall 1977 # of Students	%	A-%	Total # of Students	%	A-%
1st Semester	7	1.9	1.9	18	6.1	6.1	17	5.5	5.5	42	4.3	4.3
2nd Semester	36	9.9	11.8	31	10.7	16.8	41	13.2	18.7	108	11.2	15.5
3rd Semester	71	19.6	31.4	46	15.8	32.6	47	15.2	33.9	164	17.0	32.5
4th Semester	31	8.6	40.0	19	6.5	39.1	32	10.3	44.2	82	8.5	41.0
Graduates	174	48.0	–	136	46.7	–	120	38.7	–	430	44.7	–
Currently Enrolled or Withdrawing	43	11.8	–	41	14.1	–	53	17.1	–	137	14.2	–
After 4th Sem. TOTAL	362			291			310			963		

The study indicates that one-third of the full-time associate degree program students
never begin or complete their third semester at Toledo County Community College. Another
11 percent withdraw before completing the sophomore year. A little more than 40 percent
graduate. Approximately 14 percent prolong their education beyond the expected four
semesters and about one-third of this group graduate at a later time.

Fig. 4-7
Report with Ineffective Basic Structure

67

Page 2

The study also demonstrated the flexibility a student has in determining
a timetable for his formal education at Toledo County Community College
in that 51 (14 percent) of the 362 students sample for year 1975 were
graduated by Toledo County Community College at a time later than the
normal expected graduation date. Only five of the 51 students continued
their education on a part-time basis without interruption. Almost all
of the students who are initially full-time and realize an interruption
in their studies return to continue on a full-time basis.

In order to present the following data on attrition by program, the
three samples were combined. The total percentages do not necessarily
compute to 100 percent, because some students were currently enrolled
at the time when information for the study was collected.

Attrition by Program

Program	Percentage Attrition	Percentage Graduation
Pre-Education	67%	29%
Pre-Engineering	55%	40%
Liberal Arts	71%	23%
Liberal Arts-Math/Science	59%	39%
Business Administration	49%	30%
Accounting	71%	29%
Business Management	63%	34%
Chemical Technology	41%	33%
Drafting and Design	5%	62%
Civil and Construction Tech.	36%	63%
Data Processing Tech.	59%	40%
Electronics Tech.	51%	49%
Secretarial Science	42%	52%
Food and Service Mgmt.	29%	70%
Physical Therapy	19%	79%
Mechanical Technology	56%	43%
Criminal Justice Adm.	39%	52%
Occupational Therapy	17%	79%
Retail Management	43%	56%
Real Estate	87%	6%
General Studies	79%	11%

A second study of general interest was initiated wherein the population
consisted of all full-time and part-time students enrolled in the fall
of 1977. The random sample of 200 full-time and 200 part-time students was
used. Thus 10.6 percent of 1888 full-time students and 15.4 percent of the
1,303 part-time students were randomly selected for purposes of this second
review. Fifty-three percent of the full-time students selected were freshmen and 47 percent were sophomores, whereas 64 percent of the part-time
students were freshmen and 36 percent were sophomores.

Page 3

The attrition rate for full-time students was compatible with rates found in the first study; however, this study generated data on the part-time students not available in the first study. It was significant to note that 34 percent of the part-time students did not continue studies after the first semester and 21 percent more, for a total of 55 percent, failed to continue after the second semester. As intended, the study needs to be continued over a number of years before any further results of the part-timers can be meaningful. It was noted that 71 percent of the students listed as part-time freshmen (that is fewer than 30 total credits completed) were not enrolled four semesters later and that only 5 percent had been graduated by Toledo County Community College.

In summary, it is evident that further investigation of the enrollment trends of part-time students is very necessary since part-timers are becoming more and more an integral part of the total enrollment picture at Toledo County Community College.

Fig. 4-7
Report with Ineffective Basic Structure

A. Analyze the report shown in Fig. 4-7 and do the following exercises:

1. Does the report meet the scissors test?
2. Identify the possible audiences for the opening segment of the report.
3. Does the heading contain sufficient information?
4. Is the purpose statement adequate?
5. Is the conclusions and recommendations section sufficient?
6. Identify the possible audiences for the discussion segment of the report.
7. Would the discussion have the information all of its potential audiences would need?

B. Obtain an appropriate specimen one- or two-page report written by a technician in your field of specialization.

1. Answer 1 through 7 in A above.
2. If you have answered "no," "perhaps," or "somewhat" to 1, 3, 4, 5, or 7, explain why you think the writer did what he or she did.
3. Explain how you could edit and revise that report.
4. Edit and revise it for an effective basic structure and an effective opening segment.

C. Identify the type of generalization, if any, in each of the following sentences. Is the sentence an empirical observation, a result, a conclusion, or a recommendation?

_____ 1. Mr. Y should be provided with activities requiring gross motor movement but demanding little visual accuracy.
_____ 2. The patient is tolerating the procedure well.
_____ 3. Careless smokers cause 19.1 percent of home fires.
_____ 4. The Outstate Security System and the Detroit River Control System are experiencing operational problems.
_____ 5. The ward has, on occasion, picked up a chair and thrown it through a window at school.
_____ 6. The total daily flow should amount to 110,000 gallons per day.
_____ 7. We should purchase two IBM Selectric II typewriters.
_____ 8. The system was functional in the local manual mode and in thirteen or fourteen automatic modes.
_____ 9. The subject failed to make a reasonable contribution to the rent.
_____10. The proposed housing facility will provide apartments for 252 people.

D. Rearrange each of the following sentence outlines so that the outline goes from general to particular:

1. a. The conventional vehicle with the redesigned grill support was endurance-tested for normal urban and rural use.
b. The redesigned grill support will be satisfactory for conventional vehicles after minor modification.
c. Approve the redesigned grill support as modified for release on production vehicles.
d. The conventional vehicle with the redesigned grill support was endurance-tested for 18,760 miles.
e. The grill support required only minor adjustment during endurance testing.

The general-to-particular outline would be:
_____
_____
_____
_____
_____

2. a. Type I carbon paper is longer-lasting than Type II carbon paper.

b. Central typing should use Type I carbon paper.

c. Type I carbon paper produces ten legible carbon copies.

d. Type I carbon paper is superior to Type II carbon paper for office use.

The general-to-particular outline would be:

_____

_____

_____

_____

3. a. The photoelectric detector can be used in a tandem connection with the ionization detector.

b. The two-alarm system has good response times for both smoldering and flaming fires.

c. The estimated cost of the two-alarm system is $75.00/house.

d. A two-alarm system using both photoelectric and ionization smoke detectors should be used throughout the subdivision.

e. For the two-bedroom home test, the photoelectric detector was placed in the hallway between the bedrooms and the ionization detector was placed in the kitchen.

f. A two-alarm system using both photoelectric and ionization smoke detectors is more effective than either one-alarm system.

The general-to-particular outline would be:

_____

_____

_____

_____

_____

_____

5.1 Introduction

Georgia DeLa Cruz, a technician in the same department as Len LaFara, usually reads the discussion of every report that crosses her desk. She did very well in her technical courses and finds technical information interesting. She enjoys a good technical report even more than the crossword puzzle in the morning paper. This morning, while her co-workers are spending the first half hour doing the crossword puzzle together, she will read the report that came in yesterday afternoon. She glances at the subject and flips right to the discussion.

After reading a page, she stops for a second. What *is* this report about? The author forgot to mention that fact in the discussion. Frowning, Georgia turns back to the first page to look for the subject. Not much help there. Next, she turns to the end of the report for the conclusions. No conclusions are there. The author evidently expects her to find the conclusions sprinkled throughout the discussion, but she can't make sense of the discussion. Every third line seems to be about something different, as if the writer had put each line on a card and shuffled the deck. Often, there are references to passages, figures, and statistics on other pages. Georgia is flipping impatiently through the report, getting more and more irritated.

An hour later, she has reached the end of the report, but she still doesn't understand what the writer was saying. She should have done the crossword puzzle; at least it all fits together.

Georgia expected a structurally clear, coherent, and complete discussion, but instead she got a mess of facts. The report communicated nothing to her. Let us hope the writer was not expecting her to do something.

Although the opening segment is the most important part of your report for some audiences, the discussion segment is where you explain and justify your conclusions and recommendations. The discussion segment therefore is also very important, because in it you must convince the reader of the validity of your results and the appropriateness of your conclusions and recommendations. Your discussion reflects the quality of your work as a technician.

Unfortunately, writing the discussion is not easy. The discussion lacks the focus of the opening. We can establish fairly precise guidelines for writing the heading, the purpose statement, and the conclusions and recommendations, because the opening segment is audience-oriented. The discussion, however, consists of the particulars of your technical activity. These particulars can overwhelm a writer by their sheer bulk and number. In addition, they can be sorted in several ways, each of which might be effective for one report but ineffective for another. You will not find it easy to select and arrange the particulars appropriately.

Accordingly, the objectives of this chapter are to enable you to:
1. Understand the relationship of the discussion segment to the opening segment.
2. Select appropriate material and establish an effective structure for the discussion segment.
3. Understand the alternative patterns for particulars in the discussion.
4. Choose an appropriate pattern.

5.2 Linking the Opening to the Discussion

In the last chapter, we saw that the opening segment begins with the general and then goes on to the particular. Although it follows the opening, the discussion segment itself is not all particulars. It too has levels of generality; that is, it has its own structure. How, then, is the structure of the discussion segment connected to the structure of the opening segment? There are three types of connection: implementing a pattern, presenting background material, and amplifying the problem.

5.2.1 Implementing a Pattern

In short reports, the discussion is connected to the opening segment because it develops a pattern established in the conclusions and recommendations section. Suppose for example that the conclusions and recommendations unit has two conclusions and a recommendation. The discussion segment then would have the pattern: 1. Explanation of conclusion one; 2. Explanation of conclusion two; 3. Explanation of the recommendation. Derivation of conclusion one would be a paragraph or paragraph cluster immediately following the conclusions and recommendations.

Such a report would have the following outline:

Opening Segment	Purpose Statement
	Conclusion 1
	Conclusion 2
	Recommendation
Discussion Segment	Explanation of Conclusion 1
	Explanation of Conclusion 2
	Explanation of Recommendation

In this method of connecting the two segments, the opening segment provides an overview of the particulars in the discussion. The discussion is organized according to the pattern established by the conclusions and recommendations, without intervening material. This method often is appropriate for one- or two-page reports.

5.2.2 Presenting Background

In some reports, background material provides a transition between the opening and discussion segments. The background material either puts the technical investigation in context or presents the technical information necessary to understand the analysis in the discussion.

Putting the technical investigation in context means relating it to previous work or to an organizational problem. You sometimes must explain previous work so that the reader knows where your technical investigation started. For example, to demonstrate a need for additional staff to handle increased work loads, you might have to explain why the present staff organization was authorized. You can describe the organizational context of your technical investigation by narrating the sequence of events leading to your assignment and technical activity. For example, in proposing a safety program on protective equipment, you might start by explaining why you came to investigate the improper use of equipment in the first place.

To present the technical information necessary to understand the analysis in the discussion, you might have to explain unfamiliar technical concepts, introduce criteria and specifications, or discuss selected test results, methodology, or the equipment tested. This information often is presented in tables, as lists, or in telegraphic sentence form. Such sections of background material are set off from the discussion proper by such headings as "Test Results," "Parts Tested," etc.

This information about the technical investigation is often necessary for the reader to be able to understand the analysis of results and conclusions in the discussion. This information provides a transition between the general technical information in the conclusions and recommendations in the opening and the analysis in the discussion. It is presented separately so that you do not have to interrupt the discussion. When the reader can understand the analysis without it, this information should be presented after the analysis or in the appendices.

5.2.3 Amplifying the Problem

In many reports the discussion segment is connected to the opening segment by an amplification of the technical problem. This is essentially a development of the second element of the purpose statement in the opening segment, where you have introduced the assignment and technical tasks. In the purpose statement, you often cannot go into sufficient detail for your readers to understand the particulars of the investigation. This serves to lead into the discussion.

If your technical problem is that the drain in the parking lot is apparently plugged, for example, you may have to identify the technical issues. The first issue is the grade toward the drain in the parking lot; the second issue is the grades of the surrounding land fill; the third is the volume and pattern of water flow; the fourth is the designed capacity of the drain; and the fifth is the nature and source of the materials plugging the drain. Is the problem drain size; insufficient water flow;

excessive water flow; litter, natural leaves, and sediment; or some combination of these? All you need to state in the opening segment is that the drain in the parking lot apparently is plugged, no matter what your recommendations may be. In the beginning of the discussion, however, you have to identify the technical issues involved, so that the readers understand and accept your analysis of the problem.

5.3 Establishing the Structure of the Discussion

In the discussion, you are faced with selecting and arranging numerous ideas and details. Too often, report writers fail to be selective and thereby overwhelm their readers with particulars. Report writers also commonly launch into particulars without first establishing a structure. Before you start out, take a minute to determine what the structure of your discussion should be. First, decide what specific information can be put in appendices rather than in the discussion. Second, determine an appropriate general-to-particular structure. Third, discuss the matter in descending order of importance.

5.3.1 Selection of Material

For the discussion, you should select from your raw data and results only what you need. Do not fill your discussion with facts in hopes that "the facts speak for themselves," as so many writers are tempted to do. Unselected and unarranged information makes the writing inefficient and unclear.

You select information by referring to the specific technical issue you are addressing. For example, the issue could be whether the same correction factors apply when a new

buffered dispersing agent is used. The complete set of particulars might include the specifications of four different dispersing agents, the chemical compositions and percentages of the buffering agent in each dispersing agent, and the correction factors to be used with each dispersing agent. The issue, however, dictates that the particulars relating to correction factors are most important. Most of the other details can and should be put into appendices. The simple answer "The same correction factors do not apply for the new buffered dispersing agent" requires only certain selected information to be documented in the discussion.

An additional example follows in the same report. Because the same correction factors do not apply, new correction factors must be used to determine how much dispersing agent (salt) to use. The issue, then, is what new correction factors should be used so that future needs can be correlated with previous use. The total mass of detailed information includes calculations of chemical transfers at various temperature and hydrometer readings, yet only the new correction factors need to be presented in the discussion. Selection finally means being realistic. After all, the purpose of the report is to explain to the County Roads Commission how much of the new salt should be used on the roads for any given snowfall at any given temperature. Those readers need only to be told how to adjust the rates of application of salt to the roads.

5.3.2 General-to-Particular Order

Readers need generalizations as signposts along the route. They need an overview of the report before they can follow the details. And they work with generalizations more than with particulars. You need a basic generalization and second-level generalizations. The basic generalization usually consists of a summary paragraph either at the beginning of the discussion or between the background material and the analysis. Second-level generalizations are short paragraphs or groups of sentences introducing each part of the discussion.

For example, examine the generalizations in the following discussion segment of a parole violation hearing report.

ALLEGED VIOLATION:
Mr. John Smith, paroled on 4/15/79, allegedly violated his parole on 5/16/79 by stealing a record from the Sears store in Columbus, Ohio. Smith denies the allegation.

FINDINGS:
The board is satisfied on the evidence that Smith did violate his parole. Sears security guard T. Sawyer testified that she saw John Smith hide a record in his coat and that, when she and other employees chased Smith, she saw the record fall from his clothing. This testimony was supported by the testimony of L. Olsen, a Columbus policeman who works as a part-time security guard at the store.

Parole agent R. Young testified that after his arrest Smith admitted the incident and discussed the possibility of plea bargaining. However Smith now denies that he was even in the Sears store.

DISPOSITION:
Smith's return on warrant is approved. Parole order 4/15/79 is rescinded.

Date of violation and date returned to custody: 5/16/79.

Notice how the "Alleged Violation" section provides a basic generalization, giving the reader an overview of the issue discussed in the findings (the analysis). The first sentence of "Findings" section is a second-level generalization. This statement is the conclusion of the analysis, and it comes before the particular evidence is presented.

5.3.3 Descending Order of Importance

When you present material of about the same level of generality, you should use a descending order of importance. This is similar to the general-to-particular order; both strategies present the reader with the most important information first.

For example, if you are analyzing the effectiveness of counseling procedures, you will start with counseling by faculty rather than with the time-schedule course listings. If you are presenting testimony, you will start with the testimony of the primary witness, Ms. T. Sawyer, rather than with that of other employees and supporting witnesses. If you are itemizing job characteristics, you'll start with responsibility for supervising typists rather than with clerical duties. In each instance, the most important information for your purpose comes first.

Finally, keep your discussion of the information in proportion to its importance. Descending order of importance means that less important information receives less attention.

5.3.4 Alternative Patterns of Particulars

Introducing the discussion and establishing an effective structure are basic strategies. You now have to write the analysis. After you rough out the discussion, you still have to choose the most effective way of presenting the particulars. After all, the purposes of the analyses will vary. Sometimes you will seek to persuade, sometimes to explain, sometimes to describe, and sometimes to predict. Each of these purposes requires you to choose an appropriate pattern for the details. In order to help you sort out particulars, we outline eight alternative patterns:

Persuasion
Problem/Solution
Cause/Effect or Effect/Cause
Comparison and Contrast
Analysis
Description
Process, Causal Chain, and Instructions
Investigation

For each pattern we explain the purpose, illustrate an effective arrangement, and provide an outline.

5.3.4.1 Persuasion

You use this pattern when your primary purpose is to support or prove your conclusion or recommendation. Although most of your reports are implicitly persuasive, you often need to make the persuasion explicit. When you recommend that a new car jack not be issued with next year's model, you explicitly argue that in tests the jack failed to meet performance standards and you are implicitly arguing that the results are reliable.

Begin with a direct statement of your basic conclusion. This generalization controls the persuasive pattern. In multiparagraph segments, this will be a short core paragraph. In a single paragraph, it will be a core sentence.

Next, support your basic conclusion by presenting your reasons one by one. Put your most important point first, and explain it. Then present your other points in descending order of significance. If your argument is lengthy, each of your points will form a paragraph of its own. If your argument is short, related points can be grouped in a paragraph.

Finally, evaluate negative arguments. Are there points to be made against your conclusion? If so, why have you not accepted them? Why have you rejected them? Why do your positive points outweigh these negative points? In short, anticipate and refute the objections that can be made to your conclusion. Do not, however, spend more space refuting objections than you spend supporting your conclusion.

As an example of an effective persuasive pattern, consider the parole recommendation report in Fig. 5-1. The arrangement follows the pattern we have just presented: statement of conclusion, statements of support, refutation of objections.

The first paragraph summarizes the recommendation. The ward will be released under the social treatment plan and will attend school full-time. His positive behavior at the training school indicates that he can be rehabilitated, and this possibility outweighs the risk involved.

The next three paragraphs present the support, with the most important point presented first. The ward's group behavior indicates that he has incorporated a new value system. In addition, he has performed community service well. His health problems also should be taken into account.

The following paragraph evaluates the argument against the recommendation. The question of maturity is resolved in the ward's favor. This argument against the recommendation must be refuted if the positive argument that the ward has incorporated new values is to stand.

The final paragraph is a restatement of the basic conclusion, a common feature in persuasive reports. Notice that secondary details concerning the recommendation also are presented.

The persuasive pattern, then, has this outline:
1. Statement of conclusion.
2. Support for conclusion, arranged in descending order of importance.
3. Anticipation and refutation of objections (if any need discussion).
4. Restatement and explanation of conclusion.

The persuasive pattern should be your first choice from among these possibilities. If it does not seem right for your purpose, consider the other seven patterns.

5.3.4.2 Problem/Solution

You use this pattern when your primary purpose is to provide a solution to a problem or an answer to a question. This pattern has an element of persuasion in that you are proving that you have the solution to the problem. For example, if a new piece of test equipment must have bench space, that poses a problem for which you must provide the solution. If a fourth person is added to the office, you provide the answer to the question of whether the existing clerical staff can handle the increased work load. The problem/solution pattern is also the pattern for a proposal.

Begin with a general statement of the problem or question. This statement is important because the way you state the problem will to some extent determine the acceptability of your solution. You often even have to prove that the problem is serious enough to require a solution.

Second, establish the criteria for a solution. How do you know when you've found a solution to the problem or an answer to the question? This information often derives directly from the particulars of the problem. In this second step, you can present the details of the problem while establishing criteria for a solution.

Third, state your solution or answer. This should be a general statement showing that your solution fits the problem you've established.

Fourth, explain your solution point by point in descending order of importance. This detailed explanation supports your solution or answer.

Fifth, if it is necessary or appropriate, account for deficiencies or dismiss alternatives. In a typical problem/solution pattern, you are demonstrating here that you have found a solution, rather than choosing from among alternatives.

As an example of an effective problem/solution pattern, consider Fig. 5-2. The arrangement follows the pattern: statement of the problem, criteria for a solution, statement of the solution, and explanation of the solution. Because this is a proposal, a least-cost statement concludes the report.

The first paragraph states the problem: demand for copying services exceeds the capacity of the present copying system. This paragraph also introduces the criterion for a solution: cost.

The second paragraph states the solution, a Xerox 9200 copier.

The third and fourth paragraphs explain the solution. The primary benefits are those resulting from the automatic features of the proposed copier. Secondary benefits come from the additional services that can be provided.

The report concludes with a cost analysis. In many proposals the cost analysis is important. In this particular proposal, the cost analysis amounts to a dismissal of alternative solutions.

```
                    GENERAL MEMORANDUM
                    State of New Jersey
                 Department of Social Services

              Youth Parole and Review Board


IN THE MATTER OF:          Mark Phillips    EN2801Y
                          (Name and Number of State Ward)


DATE:          January 12, 1980

TO:            Youth Parole and Review Board

SUBJECT:       Release report and recommendation

HEARING:       January 4, 1980

PLACE:         Rosedale Training School


RECOMMENDATION:

     This report, concerning a 15-year-old state ward, Mark

Phillips, is a recommendation for the release of the ward under

the community treatment plan.  The report represents a consensus

of those present at the hearings.  The release plan involves

the return of Mark to the custody of his parents, full-time

enrollment at Morris High School, and an attempt to obtain a

part-time job.  The release is recommended on the basis of Mark's

consistent positive behavior during his stay at Rosedale Training

School.  Mark is well on his way to rehabilitation, and the risk

that he may return to delinquent activity is outweighed by Mark's

potential to become a strong, positive young man.


FINDINGS OF HEARING:

     Mark's placement at Rosedale Training School has resulted in

a positive restructuring of his values.  During his stay at the
```

Fig. 5-1
Parole Recommendation Report with Persuasive
Pattern

-2-

school, Mark has proved that he is willing to contribute positively
to his group. He also accepts help from his group. He has on
numerous occasions helped other group members with their problems,
pointed out truancies, and demonstrated good judgment and restraint
in dealing with provocations which in the past may have led him
to acts of physical violence. One incident in particular is
reported. On 6-17-79 he and two other group members worked for
a staff person who was involved in a landscaping business in
order to pay back a farmer for some equipment which had been
destroyed by a fellow group member. Mr. Perkins, Mark's group
leader, has indicated that Mark's behavior has undergone constant
improvement since his admission to the training school and is
consistently positive.

Mark's satisfactory performance in serving the community is
also of importance. Since August 20, 1979, Mark has been working
part time as kitchen help at the Parks School. This school is a
county institution for physically and mentally handicapped
children. According to the school staff, Mark has proven himself
to be an excellent worker and has been very considerate of the
schoolchildren.

During his stay at the Rosedale facility, Mark was discovered
to be suffering from chronic hypertension. This condition may
have been a contributing factor to Mark's antisocial behavior,
which led to his placement in the training school. Mark is
presently receiving medication for this ailment. The community
services worker will ensure that there will be follow-up

Fig. 5-1
Parole Recommendation Report with Persuasive
Pattern

81

appointments at City Hospital.

Mr. T. Arnold, a psychologist who examined Mark, has proposed that Mark's detention at Rosedale be continued. The psychological examination report was read into the record, and both Mark and his parents had an opportunity to review the conclusions of the psychologist. Mr. Arnold contends that Mark has only conformed his behavior to what is expected of him by the training school staff and has not fully incorporated a new value system into his mode of living. This conclusion was extensively discussed by all staff, group members, Mark, and his parents. They concurred that Mr. Arnold's conclusion was not justified in view of Mark's performance since his admission to the training school. Mark has demonstrated that he can operate very well within a group and is cooperative not only with the school staff but also with his peers. He has proved himself capable of handling the responsibility involved in his job and has received the praise of his employers. In brief, everyone who has had contact with Mark has been impressed by his performance and maturity.

Based upon the evidence for Mark's behalf, it is recommended that he be released under the community treatment plan. This plan involves Mark's return to his parents' home, full-time enrollment at Morris High School, and an attempt to obtain a part-time job. Furthermore, Mark will be required to make weekly contacts with a staff member from Rosedale Training School. Finally, the handling or possession of any weapon is strictly forbidden. This recommendation is made with the knowledge that Mark has shown consistent positive behavior from the begin-

Fig. 5-1
Parole Recommendation Report with Persuasive
Pattern

The Technician as Writer

ning of his placement at the training school. Therefore, the
risk that he may return to delinquent activity upon his release
is outweighed by the possibility of his ultimate rehabilitation.

Oliver Holmes $\mathcal{OH}$.

Presiding Officer

OH:vw

PRESENT AT HEARING: Mark Phillips, Ward

                           John Phillips, Father

                           Sally Phillips, Mother

                           Edward Williams, Group Leader

                           Dorothy Hammond, Rosedale County Community
                             Services Worker

                           Mona Walsh, Youth Specialist

                           Ben Morris, Group Member

                           John Sadecki, Group Member

                           Nate Coleman, Group Member

                           Victoria Winkler, Teacher

                           Oliver Holmes, Presiding Officer

EXHIBITS: Exhibit #1: Psychological Examination report prepared
               by T. Arnold, dated 10-18-79

          Exhibit #2: Release Readiness Report and summary pre-
               pared by Edward Williams, dated 12-20-79

          Exhibit #3: Community Treatment Plan prepared by
               Dorothy Hammond, dated 12-13-79

          Exhibit #4: Findings of Fact

Fig. 5-1
Parole Recommendation Report with Persuasive
Pattern

ANN ARBOR GENERAL HOSPITAL

Inter-Office Correspondence

UNIT	LOCATION	DATE
Administration	A6018	May 19, 1979

NAME
S. A. Kingsfield, Administrative Manager

FROM	DEPT. & LOCATION
J. C. Ruffing, Systems Analyst	Operations A5006

SUBJECT

Hospital Printing Capacity Recommendation: Xerox 9200 Copier

Purpose. The addition of the new outpatient wing to the hospital
has resulted in an increased amount of paperwork associated with
processing these additional patients. This can be met by adding
an operator during the second shift or by upgrading the copying
system. A cost analysis indicates that a new copier should be
leased.

Recommendation. To meet a projected printing requirement of
900,000 copies per year, a Xerox 9200 copier should be leased.
This copier has an automatic document feed and an automatic
collator. The total monthly cost will be $1,925.00/month. The
existing printing system with overtime added would be about
$2,465.00/month.

Benefits of the Xerox 9200 Copier. The Xerox 9200, by virtue of
its automatic document feed and automatic collating capabilities,
will provide the necessary production to meet these increased
printing demands. The automation feature of the Xerox 9200 increases production and actually lowers operating time. The
operator thus can help with the other clerical tasks part of the
time. (This advantage has not been considered in the cost
analysis.)

Additional features, which are not available on our current 1200
system, are offered on the Xerox 9200 Copier. It (1) reduces
cleaning and maintenance requirements, (2) minimizes personnel
training, (3) eliminates the master plate-maker, and (4) provides
multiple reduction capabilities.

–2–

<u>Cost Analysis</u>. The cost evaluation is based on the printing require-
ments of the fully implemented new wing, as projected over an
entire year. The requirement will be 900,000 copies per year,
which with the present system can be met only by adding an
operator during the second shift. The present JL 1200 Copier
utilizing overtime is compared with the Xerox 9200, the only
copier on the market which offers automatic document feed and
collator. These features enable the Xerox 9200 to meet the
printing requirement without an additional operator. The monthly
costs are:

	Xerox 9200	JL 1200
Leasing and Operating Cost	$1,925.00	$ 765.00
Second Shift Personnel Cost	-0-	$1,700.00
Total Monthly Cost	$1,925.00	$2,465.00

The second-shift personnel cost includes salary plus fringe
benefits.

If you have any questions or comments regarding this study, please
contact me.

Fig. 5-2
Proposal with Problem/Solution Pattern

85

Authors' Analysis of Problem/Solution Pattern

in Ann Arbor General Hospital Report

Units of Report	Problem/Solution Pattern
Purpose	1. Statement of the problem.
	2. Statement of criteria for the solution.
Recommendation	3. Statement of the solution.
Benefits of the Xerox 9200 Copier	4. Explanation of the solution point by point in descending order of importance.
Paragraph 1	Primary benefits.
Paragraph 2	Secondary benefits.
Cost Analysis	5. Dismissal of alternatives.

Note: In this particular report, the problem/
solution pattern provides the basic two-segment,
general-to-particular structure of the report.

In Fig. 5-2, as in many reports, it would seem on the surface that written communication would not be required. After all, the choice of a new copier can be handled informally and orally. In this case, however, approval of the administrative manager was required because money was involved. Any expenditure over $500 in this organization has to be approved by the executive director, whose approval is obtained through a written memorandum.

The problem/solution pattern has this outline:
1. Statement of the problem or question.
2. Statement of criteria for a solution (often combined with particulars of the problem).
3. Statement of the solution or answer.
4. Explanation of the solution point by point in descending order of importance.
5. If necessary, explanation of deficiencies or dismissal of alternatives.

5.3.4.3 Cause/Effect or Effect/Cause

You use this pattern when your primary purpose is to establish the cause of a known effect or to predict the effects of a known cause. This pattern has an element of persuasion, because you are establishing the cause or causes of an effect or asserting the effect or effects of a cause. When you explain that the frequency of accidents at an intersection results primarily from inadequate lane markings at peak traffic periods, you are establishing the cause of an effect. When you argue that a proposed rezoning will create a traffic hazard, you are predicting the effect of a cause.

Begin with a statement of the issue in question. Is it a question of cause or causes? Is it a question of effect or effects? You should immediately establish exactly what the reader should look for in the following paragraphs. You also should forecast your conclusion. State that you have analyzed three possible

causes, for example, or identified three possible effects.

Next, support your conclusion. Analyze possible causes or effects one by one. If you have several causes or effects, start with the most important. If you have several reasons for supporting a cause or an effect, start with the most important reason. In the following paragraphs, present other causes or effects or present other reasons, in descending order of significance.

Finally, anticipate and refute alternative causes or effects. You will have analyzed and dismissed these alternative possibilities in your investigation; you should account for them now. You especially must account for alternative causes or effects that are likely to come to the mind of any of your readers.

As an example of an effective effect/cause pattern, consider the discussion of the report in Fig. 5-3. The arrangement follows the pattern: statement of the issue, analysis of possible causes, anticipation of an alternative cause.

The first paragraph states the issue: protective rubber gloves are misused and are creating a safety hazard. But exactly what is causing the misuse? The causes need to be identified before a safety-training program can be established.

The second paragraph identifies and explains the primary cause of the misuse: the same rubber gloves are used for setting poles and for working on energized conductors. The same rubber gloves are used by everybody, a second cause related to the first.

The third paragraph explains the secondary cause of the misuse: the protective covering on the rubber gloves is punctured and frayed by the work aloft.

OPa Ohio Power Authority
1200 Arlington Drive
Post Office Box 1629
Columbus, Ohio 43210

SUBJECT: Misuse of protective rubber gloves DATE: August 28, 1979
         in transmission line construction.
TO: D. T. James, Director, Safety   ADDRESS: Unit B-6

SUMMARY

Maintenance Department requested that Safety investigate
complaints by linemen that our rubber work gloves posed a safety
hazard because of ineffective insulation. As part of my summer
assignment, I investigated the work situation and the use of the
gloves. This report presents the causes of the problem so that
an effective safety training program can be implemented.

I found that misuse of the gloves creates a safety hazard.
The gloves wear quickly and receive improper care. However,
ineffective company procedures rather than worker carelessness
seems to be the basic cause of the problem. These procedures
should be corrected and a safety training class conducted on
proper use of the work gloves.

INVESTIGATION

The use of rubber gloves when working on high voltage conductors is necessary to ensure the safety of the workers. These
gloves serve as an effective insulator between the conductor and
the worker. Recently linemen have complained that our current
gloves do not insulate effectively. Maintenance foremen agree
that the gloves are being used improperly. The exact causes of
the problem need to be identified before an effective safety
training program can be implemented.

I investigated the situation and found the major cause of
the problem to be that linemen do not use the rubber gloves
solely for work on the power conductors. For example, the rubber
gloves are used for setting poles in the ground. These gloves
are not intended for this type of heavy ground work and are
quickly damaged. A glove with a puncture or tear in it will not
adequately insulate the hands of a lineman when he is working on
a conductor. Furthermore, each employee is not issued his own
pair of gloves but instead uses whatever gloves are available.
The result is improper care of the gloves, because the employee
does not feel any sense of responsibility for them.

Fig. 5-3
Report of an Investigation with Effect/Cause Pattern
(Known Effect, Explained Cause)

The Technician as Writer

Although use of the gloves for ground work is the primary cause of their damage, the gloves also deteriorate with prolonged use on the power conductors. The rubber coating on the gloves becomes worn and cracked with age, and there are often punctures and tears from the rough edge of metal crossbraces. Even when properly used, the gloves eventually wear out and need to be replaced periodically.

The supervisory management assumes that the primary cause of damage to the rubber gloves and the resultant safety hazard are due to carelessness on the part of these workers. However, my analysis indicates that several basic company procedures are to blame for the problem. Instead of using one pair of gloves for all types of work, each employee should be issued two pairs of gloves to be considered as his own. The first pair would be the rubber gloves necessary for the electrical work, while the second pair would be used for heavy ground work, such as setting poles. The safety department should develop a safety training program, which will implement these changes and educate the workers about the importance of keeping their rubber gloves in good condition.

James W. Pistilli
Safety Trainee

JWP:vw

cc: J. Anderson, General Foreman, Maintenance
     P. Mikkolz, Training Programs, Safety
     J. Tracy, Supervisor, Safety

Fig. 5-3
Report of an Investigation with Effect/Cause Pattern
(Known Effect, Explained Cause)

Authors' Analysis of Effect/Cause Pattern

in Ohio Power Authority Report

Discussion Segment   Effect/Cause Pattern

INVESTIGATION

Paragraph 1   1. Statement of the issue.

Paragraphs 2 and 3   2. Support for conclusion:

                                   analysis of causes in

                                   descending order of

                                   importance.

        Paragraph 2   Primary cause

                                       Secondary cause$_1$

        Paragraph 3   Secondary cause$_2$

Paragraph 4   3. Refutation of alternative

                                   possibility (e.g., cause)

        Note: In this particular report, the effect/

        cause pattern provides the structure for

        just the discussion segment of the report.

Fig. 5-3
Report of an Investigation with Effect/Cause Pattern
(Known Effect, Explained Cause)

The Technician as Writer

The final paragraph introduces and discusses an alternative possiblity. Management had assumed that carelessness on the part of the workmen was the primary cause of the safety hazard. This cause is dismissed, as company procedures have been shown to be the primary cause.

The safety report in Fig. 5-2 actually concludes with a solution, the training program. Yet the substance of the discussion has been a causal analysis to explain the problem. Thus the effect/cause pattern is the appropriate pattern for the particulars. The issue is not what the problem is or what the solution is. The problem is known, and the solution is standard safety department procedure. The real issue is, rather, "What is the cause?"

The cause/effect pattern has this outline:
1. Statement of the issue (cause or effect) and forecast of the conclusion.
2. Support for the conclusion: analysis of causes or effects arranged in descending order of importance.
3. Anticipation and refutation of alternative possibilities.

5.3.4.4 Comparison and Contrast
You use this pattern when your primary purpose is to get your reader to accept your choice between two or more alternative objects, processes, or ideas. Your choice is based on an understanding of similarities and differences.

For our purposes here, your judgment is primarily comparative if you stress similarities and primarily contrastive if you stress differences. If you stress both, your judgment is comparative and contrastive. For example, if you are demonstrating that your firm's service is competitive with another firm's you are making a comparative judgment. If you are

demonstrating that an urban location is superior to a rural location, you probably are making a contrastive judgment. If you are demonstrating that the services of supplier X are better than those of supplier Y, you probably are comparing and contrasting to formulate your judgment.

Begin with a statement of your judgment or conclusion. Establish the relationship you are analyzing, and itemize the criteria or specific points of comparison and contrast. These will provide an outline for the remainder of the segment.

Next, present a point-by-point analysis of the similarities and the differences. These each might take a paragraph, although important points might require several. The order of these points, of course, is in descending order of significance, as determined by your conclusion.

Finally, restate your conclusion, unless the pattern is very brief. Your point-by-point comparison and contrast ends on a minor—or, at times—negative point. Therefore, summarize your comparison and contrast, and restate your conclusion.

This pattern has variations other patterns lack. Your point-by-point analysis must be appropriate to the pattern of your report.

There are three possible analyses. Let us assume that you are comparing or contrasting A with B. If your conclusion is dominately *either* comparative *or* contrastive, the appropriate order is:
Point 1 for A and B
Point 2 for A and B
. . .

Use decreasing order of significance.

If your conclusion is dominately comparative *and* contrastive, you have two alternative arrangements. If your judgment is that A is more like than unlike B, the order is:
Similarity of A and B
   Point 1 for A and B
   Point 2 for A and B
   . . .
Difference of A and B
   Point 1 for A and B
   Point 2 for A and B
   . . .

If your judgment is that A is more unlike than like B, your order is:
Difference of A and B
   . . .
Similarity of A and B
   . . .

Again, in either pattern use descending order of significance.

Notice that we suggest a point-by-point comparison, contrast, or both. This method focuses on the relationship you are analyzing and on your reasoning itself. The point-by-point method thus is more effective than the subject method, which groups all the points for A together and all the points for B together. This subject method is used only for brief, simple comparisons or contrasts.

As an example of a contrastive problem, examine the brief contrastive report on choice of carbon paper in Fig. 5-4. The arrangement follows the pattern: statement of the conclusion, point-by-point contrast, restatement of the conclusion.

The first paragraph states the conclusion. Type I carbon paper is superior to Type II carbon paper and therefore should be used. The points of comparison—durability, legibility,

erasability, and curling tendency—also are introduced.

The second paragraph then analyzes the points of contrast that support your conclusion, in descending order of significance—durability, legibility, and curling tendency.

The third paragraph analyzes erasability, the negative point of contrast, the least important to the writer's conclusion.

The fourth paragraph restates the conclusion that Type I carbon paper is superior. It also presents a summary table, which has already been interpreted.

The report concludes with a discussion of methodology. This paragraph is appendix material, because the method of the investigation need not be explained to most readers. However, in Fig. 5-4, the material is not put in an appendix, because the report is so short.

This brief report illustrates an important characteristic of the comparison and contrast pattern: it is predominately persuasive. Notice how the negative point of contrast, erasability, contains a refutation of an opposing argument: the reason for the better erasability, softness, is also a reason for poorer durability, legibility, and resistance to curling. Because you use this pattern to make a comparative evaluation, it is similar to the persuasion pattern.

The comparison and contrast pattern has this outline:
1. Statement of the conclusions and introduction of the points of comparison, contrast, or both.
2. (Comparison *or* contrast) Point-by-point comparison or contrast, in descending order of importance.

(Comparison *and* contrast, emphasizing similarities) Point-by-point comparison and contrast of similarities in descending order of importance, then comparison and contrast of differences.

(Comparison *and* contrast, emphasizing differences) Point-by-point comparison and contrast of differences, then comparison and contrast of similarities.

3. Restatement of the conclusion.

5.3.4.5 Analysis

You use this pattern when your primary purpose is to divide an object, process, or idea into its components. This pattern is primarily informative or explanatory. For example, when you explain a secretarial position to an applicant, you itemize and explain the various duties that person must perform, such as typing, filing, reception, and operating office equipment. When you explain responsibility, you break it down into the elements of supervision and evaluation of office staff, interpretive judgment, initiation of action, management of accounts, and accountability.

Begin with a summary statement of the whole object, process, or idea as the sum of its parts or components. Identify the parts or components you are going to discuss separately. This itemization outlines the following paragraphs.

Next, explain the parts one by one, grouped according to any inherent relationships between them. They should be explained either in decreasing order of significance or in some other logical order appropriate to your purpose.

Finally, conclude your analytical pattern with a restatement of the whole object, process, or idea, perhaps also interpreting it, now that the reader understands the components.

As an example of an analytical pattern, examine the occupational therapy evaluation report in Fig. 5-5. The arrangement follows the pattern: summary statement of the evaluation, point-by-point explanation of the components of the evaluation, and restatement and interpretation of the evaluation.

The first two paragraphs summarize the initial occupational-therapy evaluation, selecting a few high points from the following paragraphs. They state the objectives of the functional and diversional treatment plan and note the primary objective of keeping the patient active.

The third through seventh paragraphs discuss particular components of the evaluation: Motivation, Attitude, Mental Functioning, Physical Functioning, and Assessment of Potential. These are ordered logically, and Assessment of Potential acts as an interpretive summary for this set of paragraphs.

The eighth, ninth, and tenth paragraphs summarize Short-Term, Long-Term, and Patient's Goals. Notice that these conclusions are based upon the particular evaluations or evidence introduced in the preceding paragraphs.

The final paragraphs itemize the Treatment Plan and generally summarize or restate the evaluation. The Treatment-Plan paragraph is essentially an interpretation based on the preceding conclusions of the evaluation.

As with this occupational therapy report, analytical reports often are formatted directly. The writer itemizes and explains the material according to standard categories. Even when you are not provided with a standard format or checklist, your analytical pattern should have the same methodical approach.

GREAT LAKES STEEL

From: P J Faurl

To: A A Albright--Supervisor, Central Typing

Subject: INVESTIGATION ON THE QUALITY AND DURABILITY
OF MINI WEAR TYPES I AND II CARBON PAPER

February 15, 1980

At your request, I investigated Mini Wear Types I and II carbon paper for quality and efficiency. I tested for durability, multi-copy legibility, curling tendency, and erasability and found Type I carbon paper to be superior to Type II. I recommend that Mini Wear Type I carbon paper be used in Central Typing.

The results show the performance of both carbon types in our investigation.

1. Type I carbon paper was more durable. To test durability, we counted the number of times each sheet of carbon paper was used before a replacement was necessary. Type I was used 20 times before it was replaced, while Type II could be used only 15 times.

2. Type I carbon paper produced a greater number of legible copies than Type II produced.

3. Type I carbon paper had no tendency to curl after use, whereas Type II curled slightly.

Type I carbon paper was difficult to erase, while Type II had average erasability. Type II erased more easily because of its softness.

On the basis of these findings, I recommend that Mini Wear Type I carbon paper be used in this department. Its durability and multi-copy non-curling qualities surpass those of Mini Wear Type II. Because it is longer lasting, Type I carbon paper will be more cost-efficient.

Results of the Investigation

	No. of Times Carbon Paper Used	No. of Legible Copies Produced	Erasability	Curling
Mini Wear Type I (Hard Carbon)	20	10	Difficult	None
Mini Wear Type II (Soft Carbon)	15	8	Average	Slight

Methodology

To determine which carbon paper had better qualities, six individual transcribers were asked to use Mini Wear Type I or Mini Wear Type II carbon paper. Three used Type I and three used Type II. Each transcriber's findings were noted, logged, and averaged. Their results, based on the performance of both carbon paper types, are shown in the table.

Fig. 5-4
Investigation Report with Contrastive Pattern

95

Plymouth Rehabilitation Center
OCCUPATIONAL THERAPY

INITIAL EVALUATION

Name: Mildred De Forouw   Unit: Senior West   Date: July 19, 1979
Examiner: Lori Dostal

SUMMARY OF EVALUATIONS

The patient's initial Occupational Therapy evaluation was completed
in December 1978. The patient was found to be alert, talkative
and friendly. Her physical and mental states are fairly good,
but some assistance will be needed in some activities.

An O.T. program including ADL evaluation has been prescribed, as
well as training in precautions for maintaining safety in wheelchair transfers. In addition to these treatments, an O.T. workshop and/or floor activities will be beneficial to the patient.
The entire staff should help by encouraging her to participate in
social activities.

ANALYSIS

MOTIVATION

The patient's motivation toward socialization is good. She likes
to talk to people and be with others. Her motivation toward work
and recreational activities is fair.

ATTITUDE

The patient has a good attitude toward herself, family, peers, and
staff. She also has a good outlook on her disabilities and her
environment.

Fig. 5-5
Occupational Therapy Report with Analytic Pattern

The Technician as Writer

MENTAL FUNCTIONING

The patient's mental functioning is good, and she follows directions
well. She is alert, talkative, and cooperative. Her communication
skills are fine. She enjoys watching TV and prefers reading books
with large print.

PHYSICAL FUNCTIONING

Some assistance is needed with ambulation, as the patient's legs
are not very strong. Her upper extremities' range and strength
are within normal limits, although her neck rotation is limited.
Only minimal assistance is needed with dressing. The patient's
coordination and grasp are good, and her sensory abilities are
normal.

ASSESSMENT OF POTENTIAL

The overall assessment of her potential is fair to good because
she is cooperative and friendly. She is motivated and functions
well, but she needs persuasion to become involved in work and
activities.

GOALS

SHORT-TERM

Our goals for the patient, on a short-term basis, are for her to
establish independent activities of daily living. We also expect

Fig. 5-5
Occupational Therapy Report with Analytic Pattern

her to maintain safety in wheelchair transfer. Most important, the patient must become involved in purposeful activities around the clinic.

LONG-TERM

Long-term goals for her are to maintain the established level of functioning, which is sufficient to allow for future enrichment of the patient's life.

PATIENT'S GOALS

The patient's individual goals are to walk independently so she will not be hindered by her wheelchair. She also hopes to be able to eat in the dining room with the other residents.

TREATMENT PLAN

In order to achieve these goals, a treatment plan has been devised which involves functional and diversional Occupational Therapy. First, an ADL evaluation will be required to see if she functions normally in daily life. The patient will also need training in precautions for maintaining safety in wheelchair transfers.

To help the patient become more involved, she should participate in floor activities and/or O.T. workshop. This will also help her ambulation and motivation. Books with large print should be provided to keep her communication skills at the present level.

Fig. 5-5
Occupational Therapy Report with Analytic Pattern

The Technician as Writer

4

The treatment plan needs attention by all personnel, so I
recommend that all staff encourage the patient to attend recreational activities.

Lori Dostal

Fig. 5-5
Occupational Therapy Report with Analytic Pattern

99

The analytical pattern has this outline:

1. Summary statement of object, process, or idea to be analyzed, with identification of its components.
2. Point-by-point presentation of the components of the whole, arranged in descending order of importance or in another purposeful sequence.
3. Restatement of the object, process, or idea, with an interpretation if needed.

5.3.4.6 Description

You use this pattern when your primary purpose is to describe a thing physically, so that your readers know how it functions. You say, in effect, "Here's what it looks like and how it works." Description is similar to analysis, in that a whole is broken down into parts. Description, however, focuses on the functional relationships among the parts, rather than on the individual parts themselves. It also is restricted to physical objects, while analysis investigates any whole, such as a concept. When you describe how a dummy "survived" an impact test, you are concerned with the interaction among physical objects—the dummy, the driver's seat, the steering wheel, the windshield, etc.

Begin with an overview of the whole object's function. Explain its purpose, introduce its major components, and explain the principles of its operation.

Next, present a physical description of the object. Usually you accompany this with a figure, drawing, or layout. Identify the major features of the object, and describe its parts in a systematic sequence. This physical description is necessary for your readers to understand the functional description that follows.

Finally, present a functional description of the object. Divide the operation into its basic stages, and explain step-by-step how the parts interact, how they function together, what happens. Your step-by-step functional description should be arranged either in causal sequence or in descending order of importance. For example, if you are describing a receiver, you will begin by showing how the signal is received, then modified, and finally transmitted. A freighter design will be described in descending order of importance, from cargo capacity to power plant to arrangements and living quarters.

Do not overwhelm your readers with too many details in the descriptive pattern. Instead, always group them by parts and components, presenting them in their functional context.

As an example of a descriptive pattern, consider the separate information flyer of the report in Fig. 5-6. The arrangement follows the pattern: summary of the smoke detector's function; physical description of the object, both verbally and with a figure; functional description of the object.

The first paragraph introduces the function of the photoelectric smoke detector: to detect smoldering fires. This paragraph also explains the principle of operation, the reflection of light beams by smoke particles onto a photoelectric cell.

The second paragraph presents a brief physical description of the object. The parts of the photoelectric cell are identified in the text and figure.

The third paragraph presents a functional description of the object. In this particular case, the functional description is in a casual sequence. Smoke enters the detector and, through the sequence of events described, causes the alarm to sound.

SPENCER CONSTRUCTION COMPANY

P.O. Box 3113
Pontiac, Michigan 48214   (313) 345-8503

                                        9 September 1979

To: F. Bartell, President

From: L. Hawkins, Construction Engineer Intern   *LH*

Subject: Photoelectric Smoke Detector for Green Oaks Subdivision

     According to the Pontiac Building Code, Spencer Construction
must install one smoke detector in our Ponderosa Ranch styles
and two in our Sierra View styles. D. Finton, the Project Engineer,
has chosen the Insta-Alarum 23 photoelectric smoke detector for
its reliability and the low bid from the manufacturer's representative. I was asked to describe this detector so that our
sales agents can assure prospective clients that Spencer Construction has installed a reliable dectector that meets code
specifications.

     The Insta-Alarum 23 photoelectric smoke detector is a line-
powered unit which senses smoke and immediately sounds an alarm.
The detector is attractively housed in a plastic case, which
contains a constant light source and a photoelectric cell. When
smoke deflects the light into the photoelectric cell, the alarm
sounds. This unit has no moving parts, and it meets Building Code
specifications. The total installed cost of each unit will be
$55.00.

Fig. 5-6
Information Flyer (with Memo of Transmittal) with
Descriptive Pattern

I have prepared the attached flyer for you to distribute to the sales offices and agents for their reference. An Insta-Alarum 23 promotional brochure also is attached.

Fig. 5-6
Information Flyer (with Memo of Transmittal) with
Descriptive Pattern

SPENCER CONSTRUCTION COMPANY

P.O. Box 3113
Pontiac, Michigan 48214 ,

(313) 345-8503

THE INSTA-ALARUM 23 PHOTOELECTRIC SMOKE DETECTOR
IN THE PONDEROSA RANCH AND SIERRA VIEW HOMES

Spencer Construction Company has installed the Insta-Alarum 23 Photoelectric Smoke Detector in the Ponderosa Ranch and Sierra View homes. This detector is extremely reliable and meets all Pontiac building code specifications.

The purpose of the photoelectric smoke detector is to awaken sleeping residential occupants in the event of a fire. The most important characteristic of this detector is its ability to detect smoldering fires: the type which causes most of the deaths from residential fires. The Insta-Alarum 23 photoelectric smoke detector detects any smoke in the house and immediately sounds an alarm. The alarm activates when smoke enters a specially designed smoke chamber and scatters light from a constant light-emitting source into a photoelectric cell. This causes the photocell to trigger the alarm, which gives a continuous signal.

The photoelectric smoke detector is similar to a ceiling light fixture. It is attractively housed in a white plastic case 5 inches in diameter and 3 inches deep. Like a light fixture, it is directly connected to a conventional 120 circuit for continuous power. A schematic of the photoelectric smoke detector is shown in Figure 1. Connected to the power supply is a constant light-

Fig. 5-6
Information Flyer (with Memo of Transmittal) with
Descriptive Pattern

emitting source housed in a compartment with a convex lens. Across from the light source is a light collector. On one side of the detector are the smoke openings, through which any smoke will pass. Opposite the window, in another compartment, is a second convex lens with a photocell electrically connected to the alarm.

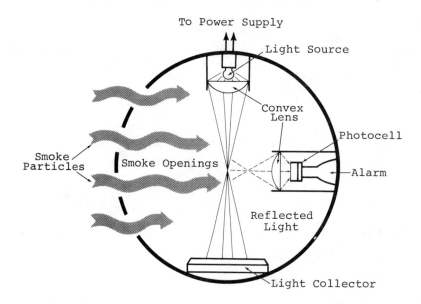

Figure 1. Schematic of Photoelectric Smoke Detector Smoke Chamber

These components of the photoelectric smoke detector inter-act either spatially or electrically. There are no moving parts. The light-emitting source is continuously powered by the house current. The light beams are refracted through the convex lens, focused through a point opposite the photocell, and then absorbed by the light collector. When smoke floats into the chamber through the smoke openings, however, the smoke particles reflect and

scatter the light beams. Because all of the light beams are
focused in front of the photocell, some of the reflected light
will enter the second convex lens and be refracted into the
photocell. This activates the solid-state circuitry in the cell,
which then triggers the alarm. The alarm immediately emits a
continuous signal with a strength of 85 decibels at 10 ft.

The Pontiac building code requires that new homes have fire
warning systems. These smoke detectors meet all the requirements
of the code. The photoelectric smoke detector has been installed
in these homes because it reacts quickly to smoldering fires.
Smoldering fires are the leading cause of death in residential
fires. Detectors which react to heat rather than smoke often
fail to awaken occupants in time. These smoke detectors have been
placed in ventilated locations close to the bedrooms. In two-
level units, one has been installed on each level. Residents may
wish to supplement this fire detection system if their house
furnishings or air conditioning systems warrant additional protection.

Fig. 5-6
Information Flyer (with Memo of Transmittal) with
Descriptive Pattern

The fourth paragraph presents miscellaneous details about the detector. These have been placed at the end so as not to interrupt the description itself. The entire pattern moves from general to particular.

Although rather simple, this example has the characteristics of the most lengthy and detailed descriptive pattern. You create an image of the details so that a reader can visualize the object and thus see how it functions. The descriptive pattern has this outline:

1. Overview of the object's function.
2. Physical description of the object, with a visual aid if possible.
3. Functional description of the object.

5.3.4.7 Process, Causal Chain, and Instructions

You use this pattern when your primary purpose is to explain a process, an event, or a set of operations that has duration in time. Your purpose is to enable your reader to understand how a result occurs or can be achieved. The process pattern is similar to that for description, but the focus is entirely different. Both process and description relate an object and an operation, or a thing and a function. Description focuses on the object, explaining how it operates so that the reader understands that object. Process, however, does just the opposite: it focuses on the action, describing objects so the reader understands the actions and events that occur. When you explain how to run a computer program, you present a sequence of actions that introduce the keyboard, visual display, and printout only as they are necessary to perform the action. You focus on the process of running the program.

The first paragraph introduces the objective of the process. You also introduce related concepts, such as the principle or method involved, and you identify and itemize the basic stages of the process. At this point, you often will use a flow chart or block diagram to illustrate these basic stages. This provides the reader with an outline of the following paragraphs.

Next, you explain the parts, equipment, or objects that are necessary for the reader to understand the process. You do this before discussing the details of the process itself, so that you don't interrupt the flow of the explanation of the process. Sometimes you will use a visual aid to explain the parts.

Finally, you explain the process stage by stage. You can present each stage in a paragraph, or you may find it more practical to group the stages within paragraphs. Introduce each stage by stating its function or output before you present the particulars. Remember that your purpose is to enable the reader to understand the flow of action or the sequence of events. Do not immerse the reader in particulars or bury the reader in details.

A final suggestion: processes, causal chains, and instructions are generally described in the order they occur in time. However, it often makes sense to divide the process into logical and functional stages, rather than into chronological stages.

As an example of a process pattern, examine the discussion segment of Fig. 5-7. The arrangement follows the pattern: statement of the objective of the process, description of the samples to be examined, and step-by-step explanation of the microscopic analysis.

The first paragraph introduces the objective of the analysis. Sixteen parts have been returned because of failures during bias-humidity aging

MURDOCK
LABORATORIES

SUBJECT: Failure Diagnostics on the 8L (F78-63)
         Case No. 38589-126
DATE: April 27, 1980   FROM: Joan Steffens
                                                 Associate Scientist
TO: A. L. Bowen, Director   Semiconductor Lab

MEMORANDUM FOR RECORD

On March 25, 1980, sixteen 8-L semiconductor packages were
returned to Murdock Laboratories after failing 80 hours' bias-
humidity aging. Upon receiving the packages, I performed ML Probe
testing to confirm and determine the mode of the failures. The
purpose of this report is to present the results of those tests
so that Quality Control can be alerted to necessary production
modifications.

Summary

Visual examination and infrared and X-ray microscopy revealed
that the bias-humidity aging failures were due to missing palladium
silicide. Oxygen contamination of the silicon surface was deemed
responsible for the prevention of the formation of palladium
silicide. Test results showed catastrophic functional failures
on twelve of the sixteen devices. Probe tests also showed no beam-
leakage failures on any of the sixteen packages. Four production
factors need correction to ensure contact window reliability.

Fig. 5-7
Investigation Report with Process Pattern

Introduction

Sixteen 8-L semiconductor packages were returned to ML Department 25 because they failed 80 hours' bias-humidity aging. Probe testing was done on two randomly selected samples to determine the cause of failure. The testing consisted of microscopic visual inspection of the metal layers and then infrared and X-ray analysis of the palladium silicide semiconductor.

The analysis of the 8-L failure was performed in three basic steps. First, the quality of the metals was evaluated through microscopic visual examination as each layer of metal was removed. Second, to determine if the palladium silicide was either poor or missing, infrared microscopy was used to show the absence of palladium. Finally, to confirm the absence of palladium silicide, X-ray analysis was done to reveal the presence of oxygen or titanium, which would indicate the lack of palladium.

Testing of 8-L

The two samples were mechanically removed from the ceramic package and mounted for visual inspection. This initial visual examination immediately revealed discoloration of metal over contact windows.

At higher magnification, missing metal and metal peeling were found, as shown in Photos 1, 2, 3, 4, 5, and 6.

To determine if the palladium silicide was either poor or missing, the metal was chemically etched away to expose the contact window areas. The metal was removed one layer at a time so that an evaluation of metal quality and integrity could be performed. The order of metal removal was gold first, followed by palladium, and finally titanium. At each step, visual examination was done and no additional problems were seen with the metals. After titanium was removed, the visual examination showed an apparent absence of palladium silicide in certain areas, as shown in Photos 7 and 8.

Confirmation of missing palladium silicide was provided by infrared microscopy, summarized in Photos 9, 10, and 11. Photo 9 shows direct visible light illumination of the suspected site, while Photo 10 shows transmitted infrared illumination of the same site. Palladium silicide attenuates infrared radiation, producing the continuous solid dark areas, while the areas of missing palladium silicide are identified by bright spots. Photo 11 provides an infrared photograph of another area.

To prove conclusively that no palladium silicide was present in certain contact window regions, an X-ray analysis was done, with the results shown in Photos 12 through 17. Photo 12 shows the window in question, as viewed under a normal microscope. Photo 13 is an enlarged X-ray scan of the suspected site. Photo 14 shows that palladium, represented by light areas, is present in some areas of the contact window and definitely missing in others. Next the probe for silicon images shows silicon everywhere except where blocked by the presence of palladium, as shown by Photo 15. Photo 16 shows oxygen present where palladium

Fig. 5-7
Investigation Report with Process Pattern

silicide was poor or missing, suggesting that a silicon dioxide layer may have prevented palladium silicide formation. Photo 17 shows titanium present in a site where no palladium silicide was found, again showing that there was no silicide formed.

These results indicate that a contact window problem caused by poor or missing palladium silicide produced the failures that were encountered after 80 hours at 85°C/85% RH bias-humidity aging. Catastrophic functional failures were confirmed on twelve of the sixteen devices. No beam failures were evidenced.

Conclusion

The results suggest that the following factors contributed to the failures:

1) The use of buffered hydrofluoric acid as an oxide etch prior to palladium deposition of silicide, leaving a residue which caused the silicide to crack and pop out.

2) The holding time between contact window sintering and evaporation of titanium and palladium was greater than 24 hours.

3) An oxide residue of either silicon dioxide or aluminum trioxide was present, because the alumina was not etched properly before the first palladium deposition.

4) Overetching of the second contact window so that sintering occurred in the entire large window area.

Correction of all four of these production factors is necessary for successful contact window reliability.

-5-

Acknowledgments

    I would like to thank R. B. Ehlers for his help and D. R. Batchelder for his excellent X-ray microprobe results and pictures.

Attachments: Photos 1 through 17

Fig. 5-7
Investigation Report with Process Pattern

tests. To determine how the parts failed, an analysis is needed, consisting of a microscopic visual inspection.

The second paragraph introduces the basic stages of the process, identified by the particular objectives of each step of the analysis, such as to determine whether the palladium silicide were either poor or missing.

The next six paragraphs present the examination process stage by stage. Initial visual examination yields some results. At higher magnification the process involves removal of layers of metal one at a time for inspection. When the core is reached, various methods of analysis reveal the absence of palladium silicide. An X-ray analysis then is conducted, confirming the absence of the palladium silicide. The writer then draws the conclusion that the lack of palladium silicide produced the bias-humidity aging test failures.

The final paragraphs present an hypothesis of why the palladium silicide is missing.

The process pattern applies to several different types of activities. It explains a process, such as admitting and examining a patient, a causal chain of events, such as an accident at a railroad crossing, or how to do something, such as how to conduct an interview with a claims applicant. The underlying pattern for each of these explanations is the same.

The process pattern has this outline:
1. Objective of the process, introduction of the method involved, and identification of the basic stages of the process.
2. Description of any parts or objects necessary to explain the process.
3. Stage-by-stage explanation of the process.

5.3.4.8 Investigation
You use this pattern when your primary purpose is to present an experimental investigation. You explain your investigation stage by stage so that your readers understand how you derived your results and accept your conclusions. The investigative pattern is similar to the one for a process, but it is a traditional pattern with unique elements. To explain the effect of the new one-way streets on traffic flow, you explain your investigation and statistical analysis. To explain the effect of stocking a lake with fish, you explain your sampling methods and subsequent results.

Begin with a statement of the purpose of the investigation, explaining precisely what was at issue and what your investigation sought to determine. This is usually a statement of the technical task or question, often including an hypothesis—what you think the answer is or what you think you're looking for.

Next, present any necessary background, the work that has already been done, previous tests, and technical background. Also explain specifications, criteria, codes, and laws that apply. Summarize the building code or EPA fuel-economy standards, for example.

Third, explain the materials and method of the investigation. The materials might be labelled "Parts Tested," and the method can be a standard test procedure. If the materials and method are unfamiliar to your readers, considerable explanation may be required. If they are familiar to your readers, this explanation can be brief, perhaps even in itemized format.

Fourth, present your results. Although your investigation will have generated many results, discuss only those necessary, perhaps presenting the full set of results in tabular form in attachments.

Fifth, analyze your results. Tell your readers how to interpret the results; don't just try to let the results speak for themselves. Your analysis of the results leads you to certain conclusions, so this analysis is central to your investigative pattern. You analyze your results to explain and support your conclusions.

Finally, discuss your conclusions and formulate your recommendations. Show how your conclusions relate to the objective of your investigation as you stated it initially. Depending on the reasons for your investigation, make recommendations for action or further investigation.

Examine the discussion segment of Fig. 5-8. The arrangement follows the pattern: statement of purpose of investigation, criteria for evaluation, materials and methods of investigation, results, analysis of results, and recommendations.

The first paragraph states the purpose of the investigation. New propellers have been suggested to replace the standard three-blade propeller currently in use. Fuel-economy gains are hypothesized on the basis of current research reported in trade journals.

The next paragraphs explain the criteria for evaluation of the new propellers, including a caution about interpreting the results. This paragraph is followed by brief sections itemizing the parts tested and the test procedures.

The next section presents the results. The results are presented in tabular form for easy reference. The technical information in the parts tested, test procedures, and test results sections provides the data base for the analysis.

Finally the results are analyzed, conclusions drawn, and recommendations formulated. The two-blade propeller for the 260 hp power plant improves the fuel-economy criterion and is acceptable under the noise criterion. The four-blade propeller, however, does not provide sufficient improvement to justify development for the 375 hp power plant. Adoption of the two-blade design is therefore recommended.

The investigative pattern has this outline:
1. Statement of the purpose of the investigation.
2. Explanation of the specifications, criteria, and technical background.
3. Itemization of materials and method of investigation.
4. Presentation of results.
5. Analysis of results and formulation of conclusions.
6. Discussion of conclusions and formulation of recommendations.

5.3.5 Choice of Pattern

Your choice from among these eight alternative patterns for arranging the particulars in the discussion segment should not be arbitrary or based on convenience. You must choose the pattern most appropriate for your purpose.

The eight patterns can be divided into two groups. Four of the patterns have a primarily persuasive or argumentative purpose, and four have a primarily explanatory or informative purpose, as the following list indicates:
Persuasive Patterns
    Persuasion
    Problem/Solution
    Cause/Effect or Effect/Cause
    Comparison and Contrast
Informative Patterns
    Analysis
    Description

form A-1048

INTER-COMMUNICATION

INLAND WATERS

Juniper Bay Test and Maintenance Facility

TO   B. Skinlea
TITLE Vice-President
DEPT. Product Development
COPIES Dept. 32 Marine Engineering
       Fuel Economy Task Force
       Consumer Products

FROM
_____
       E. J. Valery
TITLE Head
DEPT. Power Plant Testing
REF. Test Request 7932

SUBJECT
       Alternative Blade Configurations to Improve Fuel Economy

Foreward

    Inboard Panther fuel economy improvement is required to meet
Inland Water's 1982 goals. The Fuel Economy Task Force
requested vessel tests be conducted with alternative propeller
configurations to determine fuel economy improvement. Marine
Engineering therefore issued Test Request 7932 specifying the
propellers to be tested. This report presents the results of
the tests and recommendations to improve fuel economy.

Summary

    Back-to-back tests on fuel economy and noise levels were
conducted on the two-blade and four-blade alternatives
proposed to replace the standard three-blade propeller. Test results
showed improved fuel economy for the two-blade installed on the
260 hp power plant. Results also indicated no fuel economy improvement for the four-blade on the 375 hp plant. Because of improved fuel
economy as well as acceptable noise levels, the two-blade
propeller should be released on the 260 hp plant.

Fig. 5-8
Test Report with Investigative Pattern

The Technician as Writer

2

Introduction

Rising fuel costs and consequent potential customer concern necessitate evaluation of potential fuel-saving design changes for the 1982 inboard marine market. Tests were conducted with specified alternative propellers and determine the fuel-economy effect. In addition, the noise performance of these propellers was tested. Marine Engineering determined that a new two-blade design could replace the standard three-blade propeller in the 260 hp plant. They also suggested a design for the 375 hp plant. These developments indicate that the proposed configurations would improve fuel economy without significantly affecting noise levels.

Criteria for Test Evaluation

In fuel-economy testing, changes of less than three percent are deemed insignificant, because of test repeatability. In noise testing, changes of less than 0.6 dbA are considered insignificant. In all tests, the absolute numbers reported here are insignificant, since all testing was performed back-to-back.

Parts Tested

260 hp plant

Standard   12 1/4" diameter three-blade propeller

Proposed   14 5/8" diameter two-blade propeller

375 hp plant

Standard   12 1/4" diameter three-blade propeller

Proposed   10 3/16" diameter four-blade propeller

Test Procedures

Conducted back-to-back:

1. Open-water test OT-33 (16' tank)

2. Full-scale tests FT 6B (cruise) and 7H (full throttle)

3. Full-throttle noise test

Fig. 5-8
Test Report with Investigative Pattern

Test Results

Fuel Economy (Nautical Miles/Gallon)

		Tank	Water
260 hp plant	std 3-blade	38.7	31.9
	prop 2-blade	41.0	33.1
	effect	+2.3	+1.2
375 hp plant	std 3-blade	27.6	22.8
	prop 4-blade	26.8	21.7
	effect	- .8	-1.1

Noise Level (dbA corrected)

		Tank	Water
260 hp plant	std 3-blade	72.6	82.7
	prop 2-blade	72.2	82.8
	effect	- .4	+ .1
375 hp plant	std 3-blade	79.8	81.3
	prop 4-blade	80.2	81.5
	effect	+ .4	+ .2

Discussion of Results

   The two-blade propeller, design ME43-2, installed on the
260 hp plant, demonstrated fuel-economy improvement (+2.3 and
+1.2 NM/G, or 6% and 4%). Changes in the noise levels for the
two-blade propeller are within acceptable limits. The four-blade
propeller, design ME43-5, installed on the 375 hp plant, did not
demonstrate fuel-economy improvement. The four-blade did not
respond quickly to throttling in open-water testing, although no
excessive cavitation was observed in tank testing. This suggests
that the four-blade propeller did not develop the anticipated
increase in efficiency.

-4-

<u>Recommendations</u>

1. The two-blade propeller should be considered for the 260 hp engine. It improves fuel economy and does not significantly change noise levels.

2. The four-blade design should not be considered for the 375 hp engine unless the configuration can be improved noticeably.

3. Test Request 7932 is terminated with issuance of this report.

per. Alvin Karras

Fig. 5-8
Test Report with Investigative Pattern

117

Authors' Analysis of Investigation Pattern

in Inland Waters Report

Discussion Segment   Investigation Pattern

Introduction   1. Statement of purpose of

                                   the investigation.

Criteria for Test   2. Explanation of the specifi-

    Evaluation   cations, criteria, and

                                   technical background.

Parts Tested   3. Itemization of materials

Test Procedures   and method of investigation.

Test Results   4. Presentation of results.

Discussion of Results   5. Analysis of results and

                                   formulation of conclusions.

Recommendations   6. Discussion of conclusions

                                   and formulation of

                                   recommendations.

Note: In this particular report, the investigation

pattern provides the structure for just the

discussion segment of the report.

Process, Causal Chain, and Instructions
Investigation

All technical writing is implicitly persuasive, and much is also explicitly persuasive. All technical writing is done with the implicit assumption that you worked carefully and thoroughly. When you present specifications, you and your reader assume those are the correct specifications. However, technical writing is explicitly persuasive when your purpose is to effect change in an organization, to get something done.

The first four patterns are explicitly persuasive. As we outlined those patterns, you probably noticed how similar they were in structure. Each states a conclusion and then backs that conclusion with evidence. The persuasive pattern supports a conclusion; the problem/solution pattern supports a solution; the cause/effect pattern establishes a cause or predicts an effect; and the comparison and contrast pattern establishes a similarity or difference.

The four other patterns inform and explain. As we outlined those patterns, you probably noticed how the subject matter determined the structure. Each is at least as much a method of enabling the reader to understand as a way of getting someone to act. The analysis, descriptive, process, and investigative patterns focus the reader's attention on particulars so that he or she can understand the particulars.

Therefore, your purpose for presenting the particulars determines which of these patterns you will select.

Write sentence outlines for an entire report according to the following format:

Opening Segment
1. Purpose
2. Conclusions and Recommendations

Discussion Segment
3.
   3.1.
   3.2.
   . . .
4.
   4.1.
   4.2.
   . . .

Your sentences should state the point of each unit on the outline. Your teacher will show you a specimen sentence outline.

A. Persuasion versus Investigation
1. Write a sentence outline for the investigation report in Fig. 5-8.
2. Rearrange the discussion segment for a persuasion rather than an investigation pattern, and write the sentence outline for this persuasion report.
3. Explain the technique by which the discussion is connected to the opening in 1 and in 2 above.

B. Comparison and Contrast
1. Determine which of the two report outlines you have written in Exercise A would be more effective, and write a sentence outline for a report with a comparison and contrast pattern to support your conclusion.
2. Explain how the discussion is connected to the opening in this outline.

C. Analysis
Using the job classification specifications presented in Fig. 5-9 and 5-10 as a guide, analyze the nature of some actual job that is similar to these types of office positions. Write a sentence outline for an analysis report explaining the nature of that job.

D. Comparison and Contrast
Compare the job you have analyzed to the specifications listed in Fig. 5-9 and 5-10, and write a sentence outline for a comparison and contrast report arguing that the job you have analyzed in C above should be classified as C-3 or C-5.

E. Problem/Solution
Using the job specifications in either Fig. 5-9 or 5-10, establish a need. Write a sentence outline for a problem/solution report arguing that the person you observed on the job in C above provides a solution for this problem.

F. Instructions
1. Find some reasonably complex instructions for installing or using a piece of household equipment, such as installing a washer, overhauling a chain saw, preparing a three-course dinner, laying a brick patio, or assembling a sound system. Often you can find specimen instructions in do-it-yourself magazines and pamphlets. Using the process pattern, evaluate the effectiveness of these instructions.
2. Write a sentence outline for a more effective set of instructions, if you have found the instructions in 1 above ineffective.

Title   <u>Secretary</u>

FUNCTION OF CLASSIFICATION

To provide routine secretarial assistance to faculty, research and administrative staff; type reports, correspondence, and manuscripts; transcribe dictation; and to schedule appointments and receive visitors.

CHARACTERISTIC DUTIES

Type reports, memos, correspondence, forms, manuscripts, vouchers, and tabular data.

Transcribe dictation from recording equipment.

Prepare charts and graphs.

Prepare duplicating master copies.

Proofread manuscripts for grammatical and typographical errors.

Act as a receptionist by receiving visitors, answering inquiries concerning location or procedures, and scheduling appointments.

Receive and route telephone calls.

Answer inquiries from faculty, staff, students, and guests regarding applications, forms, reports, and other standard operating procedures.

Verify records, reports, and applications, checking for accuracy and completeness.

Receive, read, and interpret correspondence and determine appropriate handling.

Fig. 5-9
Job Classification Specifications for a Secretary in a
Large Organization

Operate office machines, such as calculators, adding machines,
typewriters, and dictating machines.

Process forms and applications.

RELATED DUTIES

Maintain office files and records, including the filing of
vouchers, invoices, reports, personnel records, medical
records, and financial statements.

Maintain mailing lists by performing such duties as adding or
deleting addresses.

Stuff, seal, and stamp envelopes.

Sort, collate, staple, and hand-deliver materials.

Operate office convenience unit duplicating equipment.

Perform physical inventories in unit supply rooms and post to
records.

Receive, open, and route incoming and outgoing mail.

SUPERVISION RECEIVED AND EXERCISED

Receives supervision from a designated official.

No supervisory responsibility.

Employees in this classification are expected to exercise
judgment within well-defined limits.

DESIRABLE QUALIFICATIONS

Graduation from high school, supplemented by courses in
secretarial practices, typing, and office procedures.

Ability to type with speed and accuracy.

Ability to transcribe dictation.

Fig. 5-9
Job Classification Specifications for a Secretary in a
Large Organization

The Technician as Writer

Ability to operate standard office machines.

Ability to exercise tact and discretion in relationships with
the students, faculty, and staff.

Some knowledge of general office procedures.

Reasonable knowledge of the correct usage and spelling of the
English language.

Fig. 5-9
Job Classification Specifications for a Secretary in a
Large Organization

Title Secretary Principal

BASIC FUNCTION AND RESPONSIBILITY

To provide secretarial assistance to principal faculty or
administrative staff, including relief from routine administrative details; compose interpretive correspondence;
maintain unit accounting records and procedures; adjust
faculty, student, or staff complaints; and to coordinate
unit office services.

CHARACTERISTIC DUTIES AND RESPONSIBILITIES

Collect and tabulate data for reports, records or manu-
    scripts.

Abstract information from books, journals, reports or other
    resource documents.

Prepare periodic reports of account expenditures

Maintain unit or project journal and ledger systems of
    limited complexity.

Maintain unit or project grant and salary budget accounts,
    including posting and reconciling statements.

Maintain and approve unit current account expenditures.

Monitor and maintain unit records on fellowship grants and
    traineeships, reporting irregularities.

Initiate and prepare forms and records, including payroll,
    personnel, admissions, and student grades.

Interview and perform preliminary screening of student
    applications for program entrance.

Fig. 5-10
Job Classification Specifications for a Principal
Secretary in a Large Organization

The Technician as Writer

Initiate written correspondence requiring interpretation or explanation of established policies and procedures.

Review and revise brochures, announcements, or other informational items concerning department programs.

Approve requests for materials, supplies, and services by signing requisitions or vouchers.

Answer inquiries or complaints from faculty, staff, students, or guests.

Prepare time and room schedules, course announcements, and instructional materials.

Coordinate the work assigned to assistants.

RELATED DUTIES

Type reports, manuscripts, and correspondence, incorporating specialized terminology, such as medical, legal, or mathematical.

Prepare executive and departmental meeting minutes.

Edit reports or manuscripts for organization, structure, and style.

Reconcile financial statements and balance accounts.

Receive expense statements and reconcile them with department records.

Assist in the training and orientation of new employees.

Fig. 5-10
Job Classification Specifications for a Principal
Secretary in a Large Organization

SUPERVISION RECEIVED

Receives supervision from a designated official.

Employees in this classification receive minimal supervision
and are expected to exercise judgment in the application
of policies, procedures and methods.

SUPERVISION EXERCISED

May coordinate the work of assigned assistants.

QUALIFICATIONS

Graduation from high school, supplemented by courses in
secretarial practice, typing, shorthand and office procedures, is necessary.

Reasonable progressively responsible secretarial experience
is necessary.

Experience in the performance of bookkeeping work may be
necessary.

Considerable knowledge of organizational policies and regulations.

Ability to compose and edit written materials.

Ability to establish and maintain unit bookkeeping systems,
filing and index systems, and general office procedures.

Fig. 5-10
Job Classification Specifications for a Principal
Secretary in a Large Organization

The Technician as Writer

6 Writing Sentences

6.1 Introduction

Jack Collins, a graduate of our school, is a top-notch technician who does his technical work efficiently and effectively. In most respects he knows how to write about his work, too. He can design an adequate two-segment structure for a report, he knows how to make a report move from general to particular, and he knows how to be selective in choosing what to tell his audience. Still, Jack has problems. All the way through school, Jack really never bothered with spelling, punctuation, usage, or sentence structure—all those stylistic matters teachers get so worked up about. He never thought it really counted for much. It was important to him to learn his technical subjects, but as for a few low marks in English? Well, Jack thought, nobody is perfect.

Now Jack is out on the job. He does his technical work and writes his reports, but unfortunately his problems with English are catching up with him. His supervisor has to correct Jack's reports before sending them out to his own supervisors and to contractors. Naturally, he gets angry at having to spend this time redoing Jack's reports and at being kidded by other managers who know about Jack's problem. It clearly no longer matters that Jack is a good technical man. He is costing the company time and money through his carelessness. If the economy slackens and people are laid off, Jack is likely to be among them. If someone has to go, it will be the person who gives his supervisor headaches.

Modern industry is full of people like Jack, men and women who get locked in at the bottom or who are the first to be let go when times are bad because they never learned the basic skills required to write correct English. They are generally able enough in their technical areas, but they make themselves look bad by routinely violating the conventions of language usage they should have learned in grade school or high school. True, the conventions of English punctuation are fairly arbitrary and the rules of spelling in English seem to have more exceptions than applications. The fact remains that most of us, consciously or unconsciously, fairly or unfairly, tend to think badly of people who have not mastered the basics of English usage by the time they leave school. In our eyes—and ears—people like Jack Collins call attention to themselves not because of what they can do, but because of what they *can't do.*

Our concern up to now has been with the general design of the report, getting the ideas down on paper so that they can be easily and precisely understood by the intended audience for the intended purpose. You may have wondered why we have not yet mentioned correctness in writing. After all, many people assume this is the sole requirement for clarity in a technical report. We believe, however, that the beginner must be concerned first with effectively structuring his or her thoughts. Then, later, the writer can polish, edit, and rewrite sentences and parts of sentences to make them correct, concise, and clear. Like a house builder, the writer must first do the rough carpentry, then worry about the finish work. First establish a sound structure, then worry about the trim.

Of course the truly accomplished writer, to a certain extent, at least, manages both the basic architecture and the details of appearance at the same time. Such a writer phrases even the rough-draft sentences skillfully and therefore needs to do relatively little rewriting. Most beginning writers, however, are much less successful at putting together good sentences, even if they have established a good basic structure for their reports. Their rough drafts are just rougher than those of the accomplished writer.

The purpose of this chapter, then, is to help the beginning writer gain an adequate grasp of good sentence structure. This will permit him or her to polish rough drafts and to become a skilled writer with an ear for effective English sentence structure. If you are in doubt about where to put commas, semicolons, and periods, or have never understood the structure and organization of a sentence, this explanation will prove helpful. Even if you have always dreaded English grammar, you can master at least the rudiments of clear and exact sentence structure. You will learn to recognize and control the simple sentence, the compound sentence, and the complex sentence. You will also have learned the uses of the colon and the semicolon and the punctuation of restrictive and nonrestrictive modifiers. You should then be able to:

1. Understand the different types of sentences.
2. Choose the best sentence structure and combination of sentences to suit your meaning and purpose.
3. Punctuate your sentences correctly, because you understand the structure of the sentences.
4. Edit your sentences for directness, efficiency, and clarity.

Mastery of these concepts and skills has enabled many students to improve their own sentences and punctuation. Give this chapter your close attention even if you feel rather self-assured in this department. A brief review can help even the most knowledgeable student.

6.2 A Brief Grammar Review

Because a sentence expresses a complete thought, for our purposes it can be considered the basic unit of meaning. Much in the same way that bricks and lumber are combined to build a house, these basic units are combined to produce a report. To combine sentences effectively, you need to know what a sentence is and what the various types of sentences are. In order to combine sentences for effective expression of your thoughts, you need to be able to manipulate the various parts of the sentence correctly and to use appropriate punctuation.

A brief grammar review will serve several functions. It will provide you and your teacher with a common vocabulary of grammatical terms, as well as refreshing your memory on sentence types, grammatical patterns, and punctuation practices. Furthermore, it should enable you to combine sentences effectively and appropriately in your reports.

6.2.1 The Simple Sentence

The basic type of sentence in technical and most other writing is the *simple sentence.* This can be defined as an independent clause beginning with a capital letter and ending with a period, question mark, or—very rarely in technical writing—an exclamation point. But what is an independent clause? What is a clause?

A *clause* is a grammatically related group of words consisting of a *subject* (the "performer" of the action) and a *predicate.* A predicate consists of a *verb* (a word that shows action or existence) and sometimes a *complement* (words or phrases that complete the thought). In the examples below, the subjects have been underlined once, the verbs have been

underlined twice, and the complements are in *italics:*

> Because Mr. Jones returned *to our program from the hospital on 3/12/78*
> When breaks were observed *in the circuit area*
> The patient was *alert, talkative, and cooperative.*
> Ductwork varies *throughout industrial facilities.*

The first two examples are *dependent clauses;* that is, they cannot stand alone without some additional information. The first clause tells us why something happened, but does not tell us what this was. The second clause tells us when something happened, but not what. The third and fourth examples are independent clauses; in fact, since they end with periods, they are sentences. Even if we placed periods at the end of the first two examples, they would not be sentences, because they do not express complete thoughts.

Notice that the dependent clauses would be sentences if the first word in each—a subordinating conjunction—were dropped:

> Mr. Jones returned *to our program from the hospital on 3/12/78.*
> Breaks were observed *in the circuit area.*

It is a common error to turn a good complete sentence into a dependent clause by carelessly adding or leaving out words. Sentence fragments—words, phrases, or clauses that do not express complete ideas—are to be avoided in technical writing.

Look at the verbs in these examples. The first example contains an *active verb;* that is, if you ask the question: "Who (or what) returned?" the answer is the subject, Mr. Jones. The fourth example is also an active verb: "Who (or what) varies?" Again, the answer is the subject, ductwork. The second example is not active, but *passive.* This means that the sentence does not answer the question: "Who (or what) observed?" Instead, it tells us what that person or persons observed: breaks. In technical writing, the passive verb is used to make the object being discussed the subject of the sentence, which focuses the reader's attention on that object. You should be careful when using passive verbs in technical reports, because it is very easy to leave out the responsibility for the action described in the sentence. If the reader in this case needs to know who observed the breaks in the circuit area, it would be better to write the sentence with an active verb:

I observed *breaks in the circuit area.*

The subject of a passive verb becomes the direct or indirect object of an active verb.

The verb in the third example above does not show action; it shows existence or being. The patient did not do anything; the patient was. In this case, the complement tells us how the patient was. The complement used with a being verb generally describes the subject; the complement used with an action verb generally describes the verb.

Predicate verbs show whether the sentence is in the past, present, or future. The first three examples are in past tense; the fourth is in present tense. Note that English has far more tenses than just past, present, and future; these are used to express shades of meaning, such as action completed or in progress, action of long or short duration, action before or after some other action, and so on. Consider the possible tenses that could be used in the first example:

Mr. Jones returned . . .
Mr. Jones did return . . .
Mr. Jones was returning . . .

Mr. Jones has returned . . .
Mr. Jones had returned . . .
Mr. Jones returns . . .
Mr. Jones does return . . .
Mr. Jones is returning . . .
Mr. Jones will return . . .
Mr. Jones will be returning . . .

The predicate verb not only shows tense, it also agrees with its subject. This means simply that you do not say, "Mr. Jones return" or "Mr. Jones are returning," because the subject is singular—Mr. Jones is one person. Similarly you do not say, "Breaks was observed" or "Breaks is observed," because there is more than one break, making the subject plural.

Notice that the simple sentence may have more than one subject or more than one verb. These are known as *compound subjects* and *compound verbs:*

Ductwork and wiring vary throughout industrial facilities.
Breaks were observed in the circuit area and were corrected.

Compound subjects are not separated by a comma unless the sentence has three or more in a series. In compound constructions, do not insert a comma between the last subject and the first verb:

Breaks, loose connections, and bad joints were observed in the circuit area and were corrected.

In this example, there are commas after "breaks" and "connections" because there are three subjects; there are no commas between the two verbs or between "joints" (the last subject) and "were observed" (the first verb).

So, a simple sentence is just an independent clause with some punctuation mark at the end, although this clause may have a compound subject, a compound verb, or both. You could construct your reports entirely of simple sentences, although there would be no point in doing so and too many simple sentences tend to seem choppy or boring to the reader.

6.2.2 The Compound Sentence

Once you know what independent clauses and single sentences are, you can easily recognize a *compound sentence.* To use compound sentences, however, you must understand the use of coordinating conjunctions with commas.

Just as a simple sentence is one independent clause, the compound sentence is a sentence formed by two or more independent clauses:

> These <u>standards</u> <u>can be used</u> as an important first step in defining nursing competency, but the ultimate <u>test</u> of the nurse's performance <u>is</u> in the results.
> The <u>paint</u> <u>will go</u> to the bottom, and the top <u>layer</u> <u>can be used</u> for washing brushes and rollers.

Each of these compound sentences consists of two simple sentences connected by a coordinating conjunction (and, or, but). Notice that a comma precedes the coordinating conjunction:

> The <u>ownership</u> in a corporation <u>is represented</u> by stock, and this <u>ownership</u> <u>may be transferred</u> by endorsing and delivering the stock certificate to another entity.

The compound sentence joins two closely related statements in one sentence, and the comma and coordinating conjunction expresses the precise relationship between the two statements. When the two statements are presented as two simple sentences, you lose that extra bit of precision provided by the coordinating conjunction.

If you have a question about whether you have a compound sentence or merely a simple sentence with a compound subject or verb, mentally split the sentence. If it can be divided into two independent clauses by removing the conjunction, it is a compound sentence:

> Perhaps the most common <u>uses</u> of land for recreation <u>are</u> hunting and fishing, but other <u>enterprises</u> such as camping, picnicking, and dude ranching <u>are</u> rapidly <u>increasing</u>.

> Perhaps the most common <u>uses</u> of land for recreation <u>are</u> hunting and fishing. Other <u>enterprises</u> such as camping, picnicking, and dude ranching <u>are</u> rapidly <u>increasing</u>.

Without the coordinating conjunction, each of these independent clauses can stand by itself as a simple sentence.

6.2.3 The Complex Sentence

You use the compound sentence to connect two statements, but not all statements have equal importance. The *complex sentence* subordinates one statement to another. The complex sentence consists of an independent clause and one or more subordinate (dependent) clauses.

To use a complex sentence requires you to recognize a subordinate clause and subordinating conjunctions. A subordinate clause is a dependent clause introduced by a subordinating conjunction. Here are some common subordinating conjunctions:

if	as if	after	in order to
because	although	while	until
when	even if	before	where
since	though		

A subordinating conjunction specifies the relationship between a main idea and a subordinate idea. When it introduces a construction that has a subject and a verb, you have a subordinate clause:

> When an <u>individual</u> <u>enters</u> upon the land of another . . .
> . . . although a <u>corporation</u> <u>is</u> a legal entity.
> Though the specific <u>circumstances</u> <u>might differ</u> . . .
> . . . if the <u>ownership</u> <u>is</u> widespread.

None of the above can stand by itself as a complete sentence. They are incomplete sentences, or sentence fragments. The subordinating conjunction converts an independent clause into a dependent clause that requires an independent clause to complete the meaning of the sentence.

> Simple Sentence: Only an <u>attorney</u> <u>can advise</u> you about specific situations.
> Fragment: While only an <u>attorney</u> <u>can advise</u> you about specific situations, . . .
> Complex Sentence: While only an <u>attorney</u> <u>can advise</u> you about specific situations, a general <u>partnership</u> <u>can be considered</u> as an association of two or more persons to carry on a business for a profit.

The complex sentence therefore connects two clauses together by subordinating one clause to an independent clause.

> If the <u>birds</u> <u>become</u> too hot or chilled, their <u>growth</u> <u>will be retarded</u>.
> A <u>warning</u> given to a patient too ill to comprehend the extent of danger <u>will</u> not <u>protect</u> you against liability if <u>you</u> <u>fail</u> to supervise the patient properly.

In the complex sentence, the main idea is in the independent clause and the qualifying or explanatory idea is in the subordinate clause.

As the examples show, a complex sentence can have an independent clause followed by a subordinate clause, or a subordinate clause followed by an independent clause. The order you use depends on the order of ideas in your paragraph and on the emphasis you want to achieve:

> If <u>nothing</u> <u>can be done</u> to make the condition safe, a warning <u>sign</u> <u>should be considered</u>.
> A warning <u>sign</u> <u>should be considered</u> if <u>nothing</u> <u>can be done</u> to make the condition safe.

Either order is grammatically correct.

The use of the comma in a complex sentence requires your particular attention. You need a comma after a subordinate clause that introduces a sentence, in order to indicate where the independent clause starts. When the subordinate clause follows the independent clause, the subordinating conjunction alone usually indicates the end of the independent clause and the start of the subordinate clause. The use of a comma in a complex sentence helps the reader pick out the subject and the verb of the main clause as rapidly as possible, enabling your reader to skim efficiently.

6.2.4 The Compound-Complex Sentence

The simple sentence, the compound sentence, and the complex sentence are the basic varieties of English sentences. A combination of the last two is called a *compound-complex* sentence because it contains at least two independent clauses and at least one subordinate clause. You use this combination when you have a complicated idea to express and you want to put that idea into one sentence in order to establish precise relationships. You use coordination, subordination, and punctuation to clarify these relationships.

To write compound-complex sentences, you do not need to master any new grammatical skills. Instead, you must be able to clarify the relationships among the ideas you are discussing. Here is an example of a compound-complex sentence:

It takes one week to paste student labels onto their permanent academic records, and this is very costly if no student help can be hired.

In summary, all English sentences can be divided into four categories:

Simple: Payments may be made monthly, quarterly, semiannually, or annually.
Compound: You should wrap a paint brush in foil, and then you should store it on its flat side or hang it by the handle.
Complex: When loans are repaid in a series of equal payments, the loan is "amortized."
Compound-Complex: As payments are made, the amount of principal increases and the amount of interest decreases.

Notice that the length of the sentence has nothing to do with what kind of sentence it is.

6.2.5 The Semicolon

Now that you understand the four kinds of sentences and how to join clauses together to form sentences, you need to learn the use of the semicolon [;]. Use a semicolon whenever two ideas are so closely related that they belong in a single sentence. A semicolon should be used whenever you feel that two independent clauses belong in a single sentence without the intervention of a comma and coordinating conjunction, because their subject matter is so closely related:

The upper half of the chassis showed signs of corrosion; no corrosion was seen in the lower half.
We do not anticipate problems because of uncertainty over regulations; however, it will require an effective presentation to obtain a permit for underground brine disposal.
It vaporizes easily; since its vapors are heavier

than air, it travels along the floor until it comes to a heat source; then it ignites.

As already mentioned, the semicolon must always be used to connect independent clauses when you do not use a coordinating conjunction. Independent clauses incorrectly joined only by a comma are called "run-on sentences," and incorrect use of a comma in this way is called a "comma splice" or a "comma fault." Use of either the semicolon or the coordinating conjunction with a comma corrects the run-on sentence. Either solution is preferable to splitting the run-on sentence into two simple sentences, because the run-on sentence usually results from your realization that the two ideas really belong in the same sentence.

Run-on Sentence:
The new device will handle any gauge wire in stock, it requires less manpower.

Compound Sentence with Coordinating Conjunction:
The new device will handle any gauge wire in stock, and it requires less manpower.

Compound Sentence with Semicolon:
The new device will handle any gauge wire in stock; it requires less manpower.

The semicolon is also used to join two independent clauses with a conjunctive or transitional adverb. Although the novice writer frequently neglects this, this use of the semicolon is very important because the conjunctive adverb specifies the relationship between the ideas in independent clauses far more precisely than the coordinating conjunction can.

Conjunctive adverbs show a relationship between concepts or a transition from one concept to the next. The conjunctive adverb is often used with a semicolon to join two independent clauses in a single sentence. Here is a list of conjunctive adverbs punctuated with a semicolon to construct compound sentences:

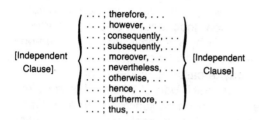

[Independent Clause] { ...; therefore, ... / ...; however, ... / ...; consequently, ... / ...; subsequently, ... / ...; moreover, ... / ...; nevertheless, ... / ...; otherwise, ... / ...; hence, ... / ...; furthermore, ... / ...; thus, ... } [Independent Clause]

As you can see, conjunctive adverbs usually are preceded by a semicolon and followed by a comma between two independent clauses.

The conjunctive adverb can be used to signal major transitions in thought, introducing a sentence rather than connecting two independent clauses to form one sentence. Major transitions occur between paragraphs and within a paragraph when a shift is made from cause to effect or comparison to contrast, for example. Use the conjunctive adverb without a semicolon in those instances:

Therefore, . . .
However, . . .

In short, the conjunctive adverb has one use, but it can be punctuated in two ways.

6.2.6 Combining Sentences

Just as a compound-complex sentence expresses the complexity of the relationships you are discussing, your combination of various sentence types in a paragraph expresses the relationships among the ideas. Let us illustrate how different combinations of sentences create different effects.

Most of the time, you can express your ideas as a series of simple sentences, as a

compound sentence, as a complex sentence, or as a compound-complex sentence. What, then, determines which type of sentence structure you use to convey your concepts? Let us look at some examples of the same ideas expressed through different types of sentences, taking note of the different effect each has. Here is a passage consisting entirely of short, simple sentences:

Insurance shifts the risk of large losses. Insurance companies are large risk bearers. The landowner should purchase insurance. First, he should consult an attorney. He should also consult a reliable insurance firm. He needs to determine what coverage, if any, is needed. Insurance shifts the risk. It does not eliminate the need to exercise care. Most insurance policies have a deductible amount. This means small losses will not be covered. The landowner will naturally wish to minimize these losses. Premium rates depend to a large extent upon the practices existing. They depend upon the conditions existing on the lands. They depend on the specific activities to be insured. These factors also affect the willingness of an insurer to protect against losses.

This series of short, simple sentences makes the ideas seem simple or childish, although they are neither, as you can see. This has the effect of writing down to the reader, who feels that the writer considers him or her simple-minded. It can also have the reverse effect, making the reader think the writer is simple-minded or has talked with too many computers. This style quickly becomes very monotonous.

Now look at this passage, which consists almost entirely of compound sentences:

Insurance shifts the risks of large losses, and the insurance companies bear the risk. The landowner should purchase insurance, but first he should consult an attorney, and he should consult a reliable insurance firm. He needs to determine what coverage is needed. Insurance shifts the risk, but it does not eliminate the need to exercise care. Most insurance policies have a deductible amount, and this means small losses will not be covered. The landowner will naturally wish to minimize these losses. Premium rates depend to a large extent on the practices existing on the lands, and they depend on the activities to be insured. These factors also affect the willingness of the insurer to protect against losses.

As you can see, a long passage of compound sentences begins to express the sophistication of the ideas, but very quickly it, too, becomes monotonous.

Now consider this passage. It mixes compound and complex sentences, has a long compound-complex sentence with a semicolon, and has a simple sentence with a compound subject:

Insurance shifts the risk of large losses to a professional risk bearer. Before purchasing insurance, the landowner should consult an attorney and a reliable insurance firm to determine what coverage, if any, is needed. While insurance shifts the risk, it does not eliminate the need for the landowner to exercise the care he ordinarily would; most insurance policies will have a deductible amount whereby small losses will not be covered, and the landowner will naturally wish to minimize these losses where possible. In addition, both the premium rates and the willingness of an insurer to protect against losses will depend to a large extent upon the practices and conditions existing on the lands and upon the specific activities to be insured.

Although this combination of sentences expresses the complicated relationship of ideas, it becomes ponderous when the semicolon joins four clauses in a forty-eight-word sentence.

Now evaluate this passage, which combines a variety of sentence types and expresses the complicated relationship of ideas even more effectively than the previous passage:

Because insurance shifts the risk of large losses to a professional risk bearer, the landowner should determine what coverage, if any, he needs. But before buying insurance, the landowner should consult an attorney and a reliable insurance firm. While insurance shifts the risk, it does not eliminate the need for the landowner to exercise care. Most insurance policies require a deductible amount to be paid by the policy holder to cover small losses, and the landowner will naturally wish to minimize these costs where possible. In addition, the landowner should understand that premium rates and even the willingness of an insurer to provide coverage will depend upon the practices and conditions existing on the lands and upon the specific activities to be insured. For these reasons, the prospective insurance purchaser should seek professional advice before buying.

This passage has stylistic variety without losing focus on the important subjects and verbs and without violating the relationships among the ideas.

The complex sentence is often more versatile than the compound sentence in expressing complicated ideas and relationships. Beginners in a language start out by expressing their ideas as simple sentences; then they begin to use compound sentences. They move on to complex sentences only as they become aware of and wish to express more precise relationships and connections between ideas. Then they combine all of these sentence types in extended passages that express exactly what they mean.

6.2.7 Other Punctuation

Although we do not intend to present you with a complete review of grammar and mechanics, much of which you already know, we would like to discuss briefly some additional punctuation techniques that have troubled at least some students.

Probably the most important punctuation problem is that of punctuation with restrictive and nonrestrictive modifiers. Failure to punctuate a modifier correctly changes the meaning of the sentence.

6.2.7.1 Restrictive and Nonrestrictive Modifiers.

A modifier describes a characteristic of a noun or pronoun. A pronoun, of course, refers back to a noun. Here are some examples, with the modifiers in *italics*. Notice that the modifier can be a clause or even a phrase:

> The debeaking of the pullets *that have cannibalistic tendencies* should be done immediately.
> The price of land *which appeals to consumers for leisure-time use* has risen 30% in ten years.
> *Considering all the information carefully,* I concluded that the second plan was the best of the three.
> The corporation, *which is not affected by the death of its members*, is preferable to a partnership in certain types of situations.

The first two sentences contain restrictive modifiers; the last two sentences contain nonrestrictive modifiers.

A *restrictive modifier* is one that restricts, limits, changes, or affects the meaning of the sentence in some conclusive or absolute way, so that the sentence would not convey the same meaning at all if the modifier were omitted.

> The debeaking of the pullets *that have cannibalistic tendencies* should be done immediately. [The pullets that do not have

cannibalistic tendencies need not be debeaked, or at least not immediately.]
> The price of land *which appeals to consumers for leisure-time use* has risen 30% in ten years. [The price of land which does not appeal to consumers for leisure-time use is not mentioned.]

The restrictive modifier is not set off from the rest of the sentence by commas.

A nonrestrictive modifier does not restrict, change, or limit the basic meaning of the sentence; instead, it simply supplies additional but unessential information. That is, the nonrestrictive modifier can be omitted from the sentence without any loss of the meaning in the sentence. We indicate this by setting the nonrestrictive modifier off from the rest of the sentence with commas:

> *Considering all the information carefully,* I concluded that the second plan was the best of the three.
> The corporation, *which is not affected by the death of its members*, is preferable to a partnership in certain types of situations.

This use of commas is important to remember, since the presence or absence of commas around the modifier can make a difference in the meaning you convey to the reader:

> The rabbit hutch, *which is open-air and self-cleaning*, has only a single deck or story. [Only one rabbit hutch has been mentioned; it happens to be open-air and self-cleaning; it has only a single deck or story.]
> The rabbit hutch *which is open-air and self-cleaning* has only a single deck or story. [This implies that there are other rabbit hutches under discussion; only this one is open-air and self-cleaning; this one has a single deck or story; how many decks the others have is not stated.]

Be sure to keep these principles in mind in writing your own sentences, because misunderstanding the nature of nonrestrictive modifiers can lead to the writing of fragments. The student who has been trained to recognize a sentence as a completed idea sometimes places the period just before the nonrestrictive modifier and, not knowing what else to do with it, incorrectly punctuates the modifier as a separate sentence.

6.2.7.2 The Colon

The colon (:) is completely different from the semicolon. They almost never can be interchanged and therefore should not be confused. The semicolon is an essential tool; the colon is an auxiliary, special-purpose tool. The primary use of the colon in technical reports is to introduce lists, examples, passages of particulars, tables and figures, and heading items.

An independent clause followed by a colon is the best way to introduce a long series of similar concepts. This provides maximum clarity. Here are some examples:

> The topics to be considered are the following: cleaning and disinfecting the house and equipment, preparing for the chicks, feather picking and cannibalism, vaccination and parasites, and proper handling and transportation methods.

> The study included three terminals:
> the Hazeltine 2000
> the IBM 1050
> the Univac 9300

> To: Arthur Leach, Production Foreman

Notice that in the report text itself (first and second examples above) a complete independent clause always precedes the colon; the series after the colon cannot act as objects of the verb of the clause that precedes the colon. This is a frequent mistake in the use of

the colon. Ask yourself if you could replace the colon with a period. If not, the colon is incorrect because it provides as definite a closing to the preceding part of the sentence as would a period. Another frequent mistake is to use a semicolon in place of the colon. Only an independent clause can follow the semicolon.

It is also sometimes possible to use the colon where a semicolon normally would be used, to tie together two independent clauses so closely united in idea that the writer feels they belong in the same sentence. The semicolon and colon function in different ways. The semicolon is virtually a full-stop connector; the colon points forward, indicating that whatever follows it resolves, completes, or explains what preceded it. This use of the colon is much like its other use:

> Mr. Yates has not consistently attended O.T. sessions for the past two months: he has either refused to come in or has attended some other activity, except for a period of three weeks.

We mention this use of the colon only in passing, because you need above all to learn the use of the semicolon. Do not confuse these two types of punctuation.

The colon is occasionally used to introduce a short word or phrase after an independent clause. This word or phrase resolves or explains the preceding independent clause, emphasizing a point:

> A knowledge of the techniques of exterior painting saves you the essentials: time, money, and energy.

6.2.7.3 Quotation Marks

Quotation marks are rarely used in technical reports, but occasionally you must quote

? what about "apple" and "oranges".

someone else. The quotation marks are normally put before and after the words of the speaker, and commas separate the quoted material from everything else in the sentence.

Mr. Talbot said, "The SOAP system of charting brings helpful information to all concerned with patient care."

"The SOAP system of charting brings helpful information to all concerned with patient care," Mr. Talbot said.

The comma and period always go inside the quotation marks.

6.2.7.4 Serial Commas

Series are common in technical reports but are simple to punctuate. Just put a comma after each item but the last in the series. To avoid ambiguity, put a comma before the "and" or "or" that precedes the last item. The series can be made up of words, phrases, and even clauses:

After the final draft is typed, check for typographical errors, grammatical structure, spelling, etc. The nurse should avoid descriptions such as "a large amount," "more," or "appears to be."

Items in a series must be consistent—that is, they must have the same form. At times you use semicolons instead of commas to separate the items. You do this when one or more items contain internal commas, which would confuse the punctuation of the series:

The release plan involves the return of the boy to the custody of his parents, who have agreed to the conditions specified above; full-time enrollment in a high school; and an attempt to obtain a part-time job.

6.2.7.5 Introductory Phrases

Introductory phrases should be separated from

the beginning of the main clause of a sentence. Keep the needs of the reader in mind. The reader must be able to pick out the subject and indicative verb of the independent clause as rapidly as possible in order to get the exact meaning of the sentence without having to reread. The comma signals the beginning of the independent clause.

Punctuation conveys part of the meaning of the sentence. The writer must use commas and other punctuation accurately in order to convey information exactly and clearly to the reader. This is absolutely essential in technical writing, where misinformation, misinterpretation, or vagueness of meaning may be very costly indeed.

6.3 Grammar Review Workshop

This is a series of exercises to test your understanding of the grammar review, with the answers at the end of this chapter. Do the exercises and then check the answers to spot your own grammar and punctuation problems, if you have any.

6.3.1 Workshop on Different Sentence Types

Underline the subjects and double underline the indicative verbs in each clause in the following sentences; then identify which type of sentence each is:

1. When instructors submit the Opscan sheets late or with errors, the system fails to operate.
2. If the landowner is considering a camp ground, he will find that some campers prefer a primitive site, but he should also realize that most prefer modern conveniences such as flush toilets, electric lights, and appliance hookups.
3. This, however, requires additional time.

4. Problems will have to be solved if the cumulative grading report is to be implemented.
5. The problem-oriented health record is a method of documenting health care to promote continuity of patient care among the various members of the health team.
6. All notes are signed with each individual's specific credentials, followed by "Respiratory Therapy Department."
7. When the bend is made in the plane perpendicular to that of the leads, make the bend at least 1/8 inch from the plastic case.
8. After interviewing those persons directly associated with Community Services, we have determined that the actual cause of failure of the system is the gathering of student information.
9. Our study to review the adequacy of the present system began in May 1972 and has just recently been completed.
10. A system can function in the manager's absence, but each trainee must be carefully supervised.

6.3.2 Workshop on Sentence Punctuation

Supply the punctuation missing in the following sentences. Some sentences need several different types of punctuation; one sentence is correct as it stands:
1. The boy who had been committed for larceny assault and truancy was not very cooperative.
2. If a trust has been established then the corporate trustee provides the level of management competency as well as the continuity.
3. Some grains such as oats barley wheat or sorghum grains may be fed but they generally present the problem of an inadequate protein intake.

4. After the chicks are six weeks old heat seldom is required.
5. The doctor bases his diagnosis on the nurses's notes therefore these must be accurate and complete.
6. Many nurses are under the impression that nurses' notes are not admissible legal evidence this is not true.
7. The circuit that is connected to the demonstration board should be tested first.
8. External parasites most likely to attack the broilers are blue bugs lice mites and fleas.
9. Chemical biological and cultural controls will be studied.
10. His reaction crying was most unusual for an adolescent.
11. The patient already anxious and depressed will not react well to questions put in a sharp tone of voice.
12. The problem list now becomes the first page of the record an index or table of contents to the record.
13. Raising rabbits pays off well especially when proper sanitary precautions are taken.
14. The ward was committed for the following offenses larceny assault and battery and truancy.

6.4 Editing Sentences

So far in this chapter, the emphasis has been upon helping you to understand how to manipulate sentences to achieve grammatical correctness. Along the way, we have introduced other considerations as well. Now, however, we want to discuss briefly the stylistic features desirable in good technical prose. Of course, correctness is fundamental and necessary, but good technical prose is more than just correct; it is direct, economical, and clear. Our concern in the remaining part of this chapter is with editing rough-draft sentences to achieve directness, efficiency, and clarity.

6.4.1 Editing for Directness

In general, you can achieve directness in English in three ways: first, by using substantive words in the subject slots of your sentences; second, by using strong, active verbs in your verb slots; and third, by following conventional English word order.

6.4.1.1 Using the Right Subject

Of course, every sentence has a subject, usually a noun or pronoun appearing near the beginning of the sentence. Indirect sentences frequently result from using nonsubstantive words instead of substantive ones. That is, writers frequently drop the real subject of the sentence and fill the subject slot with what has been called grammatical garbage. For example, someone might write:

> It is the responsibility of the Department of Highways and Traffic to investigate all fatal automobile accident sites in the city.

The real subject here is neither the pronoun "it" nor the noun "responsibility," although these words are grammatically correct as subjects. Rather the real subject of discussion is "Department of Highways and Traffic." When the real subject and the grammatical subject are the same, the sentence is much more direct:

> The Department of Highways and Traffic is responsible for investigating all fatal automobile accident sites in the city.

Similarly:

> Selection of the girder size and shape is determined by satisfying specifications, codes, and budget constraints.

Here the noun "selection" is the grammatical subject; however, common sense suggests that the real subject of the sentence is "girder." "Selection" is actually a verb idea which has been expressed in noun form and thrust into the subject slot. The sentence might be rewritten more directly:

> Properly sized and shaped girders are selected on the basis of specifications, codes, and budget constraints.

While this revision eliminates only one word, the sentence is improved because the subject ("girders") is emphasized and the action ("are selected") is transformed from a noun into the much more appropriate verb.

The same pattern is recognizable in this sentence:

> Reduction in production variance would be the effect of this change.

Here the noun "reduction" is a verb idea masquerading as a subject noun. The real subject of the sentence is "change." The sentence should be rewritten:

> This change would reduce production variance.

Not only does this change cut the sentence from eleven words to six, it focuses attention on the real subject ("change"). Moreover, it eliminates the very awkward phrase "reduction in production." (Say that phrase out loud; it sounds like a song title, doesn't it?) Reading the rough draft aloud can prevent such unconscious stylistic awkwardness.

As you can see from these examples, writers are sometimes careless with their subjects. In rough drafts, writers sometimes inadvertently replace the real subjects with garbage, often verbs in noun form. A simple but effective editorial act is to examine the subject slots in all of your sentences carefully. Have you filled them with substantive words that mean the things or ideas actually being talked about, or have you substituted something else? A quick check often reveals needlessly indirect sentences.

6.4.1.2 Using the Right Verb

Let us get rid of a myth. Not all verbs need to be active. Passive verbs have perfectly legitimate uses in the language. They allow us to maintain subject focus, for example, and to focus on the consequence of an action rather than upon the source of that action. Consider this example:

> John's arm was broken in a skiing accident.

There is nothing wrong with that sentence. We are talking about John's arm, not about the accident, so it would really be indirect to write:

> A skiing accident broke John's arm.

Of course we could say, "John broke his arm" (active verb), but that may not be desirable, because it shifts the focus from John's arm to John himself. Depending upon the context, that may not be appropriate.

Passive verbs are often useful, but many writers nonetheless rely much too much on the passive voice, forcing themselves into awkward and indirect constructions. For example, consider this sentence:

> On October 1, the Industrial Safety department was informed by you that operators of the recently purchased bag-feed machine were being injured on the job.

Here the verb is an extremely indirect passive construction ("was informed by you"). It makes much more sense to turn that into an active verb, revising the sentence to read:

> On October 1, you informed the Industrial Safety department . . .

Notice, however, that the passive verb "were being injured" is as appropriate here as an active construction and can be left as it is.

Similarly:

> This approach can be used by the Data Translator to place record instances in the data base.

By using the real subject and making the verb active, we can make the sentence more direct:

> The Data Translator can use this approach to place record instances in the data base.

You might also write:

> It is realized that any definite conclusions from these data are risky because of the small sample.

You should have written:

> I realize that these data cannot support definite conclusions because . . .

Or you might drop both yourself and "realize" from the sentence, writing:

> These data cannot support definite conclusions because . . .

In either revision you will have eliminated the very indirect passive construction "It is realized that."

You might write:

> It would be very much appreciated if you could tell me what you have achieved after implementing these suggestions.

You should have written:

> I would appreciate your telling me what you have achieved . . .

Or, to be even more direct, drop yourself out and write:

> Please report on what you have achieved . . .

Again, either revision eliminates the very indirect construction "It would be appreciated."

The examples make clear that we often use needlessly indirect passive verb constructions. We translate active verb ideas into noun form ("conclude" becomes "conclusion" as in "the conclusion was made") and thus force ourselves to replace a good active verb with an indirect passive construction ("was made"). Usually it is much more direct to use the active verb ("we conclude"). Remember of course that passive verbs have their uses, but also remember that many technical writers depend excessively upon indirect passive constructions. Don't be among them. Simply scan your rough draft to see whether the passive verb constructions you have used are necessary in context. If they are not, revise the sentence, using active verbs.

6.4.1.3 Using Conventional Word Order

English is a word-order language. That is, the grammatical function of words in sentences depends heavily upon their order, more so than in languages that signal the grammatical function with word endings. There is a definite conventional pattern for English sentences: subject—verb—object. If the word order is eccentric, the meaning may suffer and the readers may be slowed down. Consider this example:

> Microprocessor-controlled equipment will provide the necessary quality unquestionably.

Tacked onto the end of the sentence, "unquestionably" changes the meaning. The proper and conventional order for the sentence is:

> Microprocessor-controlled equipment unquestionably will provide the necessary quality.

Here the adverb "unquestionably" appears where we expect it to appear, immediately next to the verb it modifies. It could also be put at the beginning of the sentence or after "will," but at the beginning it might seem to modify "microprocessor-controlled equipment." That obscures the meaning.

Similarly you might write:

> Also by applying the adhesive to the shoe and pad bonding surfaces this process can be made cost-effective in production.

Here the subject is the fourteenth word of the sentence. Furthermore, the sentence seems to say that the process is doing the applying, when that obviously isn't the case. The sentence should therefore be recast in conventional order, eliminating the unnecessary "in production" and the passive "can be made":

> This process can be cost-effective if adhesive is applied to both the shoe and pad bonding surfaces.

You might write:

> The purpose of this report is to suggest a method of surveying cars in the field over a ten-year life span which will provide the necessary data to confirm APG testing.

Because "method" and "which" are separated by twelve words, "which" illogically seems to modify "life span." Of course, we know that can't be the case, because it doesn't make sense. But since that is what the sentence seems to say, the reader has to go back and read it again.

The same problem is evident in this sentence:

> This plan is a good one, but no procedures for the finishing of parts that are cost-effective exist at the present time.

Because "procedures" and "that" are improperly separated, the sentence seems to mention cost-effective parts. Actually it means that no cost-effective procedures exist. As you can see, a simple change in word order considerably affects the directness of the sentence:

> The plan is a good one, but no cost-effective procedures for the finishing of parts exist at the present time.

It would be even better to simplify the sentence:

> The plan is good, but no cost-effective procedures exist for finishing parts.

Not only does that change clarify the meaning of the sentence, it cuts out ten of the original twenty-three words in the sentence.

Since word order affects meaning, unconventional word order is likely to obscure or even change the meaning. Sometimes the only effect of unconventional word order will be awkwardness: the sentence just won't sound right. Still, the most important point is that word order affects meaning. Try this little exercise. See how many meanings you can get by inserting the word "only" into the sentence "he said she loves me." As you edit rough drafts, keep in mind how much these small shifts in word order can change the meaning. Try reading your rough drafts out loud; you may hear unconventional patterns more readily than you will see them.

6.4.2 Editing for Efficiency

An efficient device extracts the maximum useful work from an input of energy. An efficient automobile engine, for example, may get twenty-five miles per gallon, while an inefficient engine may get only eighteen miles per gallon: from the same amount of fuel, it can get different outputs of work.

The same is true of prose. When you write a sentence, you are building a device to do work, in this case to convey a certain meaning. If your design is good, it can do a lot of work with little input of energy (yours and the reader's). If your design is poor, the work requires a higher expenditure of energy. The work gets done in either case, but there may be considerable differences in how it gets done.

Rough-draft sentences frequently are inefficient. What might be expressed with one word gets expressed with two words or a phrase; what might be a phrase becomes a clause; a clause becomes a sentence; a sentence becomes a paragraph; and so on. The task of the writer is to shorten and simplify wherever possible. That doesn't mean short sentences are necessarily better than long ones; it means that tight, frugal sentences are better than flabby, wasteful ones, whatever their length. Consider these examples:

> I am writing to inquire whether or not there will be any openings in the near future in which I may be employed.

Cut down to say what it really says, this sentence becomes:

> I am writing to ask if there will soon be openings for which I qualify.

Or:

Are there likely to be openings for me in the near future?

Or:

Will there be openings for me soon?

Consider this example:

Also, the mean age of the manned trackers was lower than the age of the unmanned trackers. The mean age of the unmanned trackers was 37.6 years, and the mean age of the manned trackers was 36.1 years.

Here two sentences carry a one-sentence idea, as the following revision indicates:

The mean age of manned trackers was 36.1 years, of unmanned trackers 37.6 years.

The rewrite saves twenty-four of the original thirty-eight words. The original sentence was, in a sense, approximately 30 percent efficient, not very impressive for an engine or for a sentence. Look at this example:

By considering characteristic output curves of each pump, the conclusion was reached that Pump Number Two can do the work with less power input and therefore with more economical operation cost than Pump Number One.

Written more efficiently, the sentence becomes:

Characteristic output curves show that Pump Number Two can do the work more efficiently and economically than Pump Number One.

Pump number one appears to be inefficient and costly, and sentence number one seems to have the same problem. The first sentence has thirty-five words; the second twenty-one. Still, they do the same amount of work. Again, the first version is not very impressive.

It would be easy to keep giving examples of inefficient prose. To do so, however, would itself be inefficient, since the point is already obvious: words do work; efficient use of words gets the maximum possible work from every word.

6.4.3 Editing for Clarity

If more than one meaning can be read into what you write, your prose is not clear, and of course unclear technical prose is wholly unacceptable. You can easily understand why if you imagine yourself following a technical manual's instructions on making nitroglycerin or on defusing a bomb.

Two recurrent sources of unclear constructions in technical prose are ambiguous pronoun references and dangling modifiers, problems which have shown up in several previous examples. To improve the clarity of your prose, make sure you have not let either of these unclear constructions creep in.

Rough drafts usually contain many unclear pronoun references—excessive use of "its," "this," "that," and "which," many with unclear antecedents. For example:

Another picture is taken of the next picture on the the filmstrip by moving it to the next picture.

"It" appears to refer to "filmstrip." Actually it refers to a word in the previous sentence, "camera." Would you have moved the filmstrip or the camera? There is an obvious difference. Here is another example:

First, it was difficult to insulate the sample enough to prevent arcing to the wall of the oven. This was corrected by placing glass sheets on the surfaces inside the oven. Second, the insulator resistance was low enough to allow enough current to activate the "fault" detection circuit at the higher voltage. This required that the test be restarted.

What does the word "this" refer to in the second and fourth sentences?

Or:

They are responsible for running these tests and reporting on their day-to-day operation.

Whose day-to-day operation is to be reported on? Does "their" refer to "tests" or to "they"? The same ambiguity is evident in this example:

When the two tests are run on ionization detectors and photoelectric detectors, a clear-cut choice is still difficult because they are so similar.

Are the detectors similar or are the tests similar? You can't tell. Common sense says that the issue probably is which detector is the best choice and that therefore the antecedent of "they" is "detectors." Still, there is clearly no grammatical evidence to back that up, and the sentence remains unclear. To interpret it one way or the other, the reader would have to rely upon intuition. With technical prose, as we have said, that is not safe.

Dangling modifiers are also a problem in most first drafts. A dangling modifier either has no noun to modify or modifies the wrong noun. For example, the modifier in the following example is not clear:

In observing the actions of Operator Two, there was a definite idle time in each cycle.

Who was idle? You can't tell, because the phrase "in observing the actions of Operator Two" is butted up against the word "there," instead of standing next to either possible antecedent, "the operator" or "the observer," which are completely left out of the main clause.

As this example suggests, sentences which begin with participial phrases ("In observing," "Having completed," "Running down the ramp," etc.) are frequently unclear because the first noun after the phrase is the wrong noun, often because the correct noun has been omitted. Just imagine a noun slot and a verb slot to be filled after the participial phrase:

In observing the actions of Operator Two, [noun] [verb] . . .

In this case, if the noun is not "the observer" or some other noun that means the same thing, the sentence will be unclear, because the participial phrase dangles. This frequently happens because the writer has used an inappropriate passive verb. For example:

After examining a graph of the three-month average, the conclusion was made . . . .

Here it appears that the conclusion examined the graph. Yet obviously some person was doing the examination. By switching into the passive voice and dropping the real subject out of the sentence, the writer of this first-draft sentence made the result of the action rather than the source of the action the subject. The modifier dangles and the sentence is unclear. Here are additional examples:

By using a faster strain rate, viscous flow is not allowed to take place.

Is the viscous flow using a faster strain rate?

By using the following four steps as a guide, hydrogen embrittlement during plating can be minimized.

Is hydrogen embrittlement using the four steps?

Upon examination of the problem, two questions need to be answered.

Have the questions examined the problem?

Most of the time, of course, these minor ambiguities are merely annoying or absurd and cause no real uncertainty. However, sometimes they really perplex the readers. Consider this example:

> Having completed a trial run on 4% of the job, a cost and time estimate can be made to an accuracy of plus or minus 10%. These estimates are all-inclusive, assuming the only yield to be the proposed system (including tags) in a fully operational state. Using the same four warehousemen as in the trial run, seven weeks is the best estimate of the time requirement. Based on current prices, $12,000 is the estimated cost. Through inauguration of the new floor plan in conjunction with its complementary record-keeping system, worker efficiency should attain or exceed 100%.

Who completed the trial run? Who assumes? Who uses? Who inaugurates?

This example illustrates how easy it is for a writer to fall into this trap. One dangling modifier invites another. Of course, most of the time, ambiguous pronouns and dangling modifiers are more awkward than unclear. In any cluster of such sentences, however, the small increments of uncertainty accumulate, forcing the reader to stumble, slow down, and perhaps stop.

6.5 Workshop in Editing Sentences

As a test of your understanding of single-sentence editing, do the following exercise. The answers are provided at the end of the chapter. Check your own answers, then go back and review parts of the chapter which discuss errors you have made.

6.5.1 Workshop on Directness

Give the following sentences correct punctuation, substantive subjects, strong active verbs, and conventional word order.

1. The disturbed youth was engaged in many weird or crazy activities, huddling on the floor, crying, chewing on his comb, and crawling on the floor.
2. The defendant's father died in 1971 at the age of fifty-six from emphysema.
3. In checking the records at Jones High School, it was learned that the defendant received four Es, one C, and one B during his final semester at school.
4. In this case there was a severe lack of understanding of the use of wire rope clips on the part of the rigging crew involved.
5. The new fuel pumps will be used in 50,000-mile endurance testing along with test-stand endurance testing.
6. It is this manual method, and the mass flow values it provides, that is causing the reliability problem with test results, and extra costs to the research center.
7. These patchings were the result of pinhole flaws in the glass lining.
8. The Atomic Energy Commission has sent a request for additional information on the application for a reactor construction permit.
9. It is therefore seen that by using the sieve-tray distillation tower, as its operation has been explained, Midwestern Chemical will be able to save a sum in excess of $107,000 per year.
10. One other factor that may partly account for the better fuel economy showing of the diesel over the gasoline engine is the diesel's lower power-to-weight ratio.

6.5.2 Workshop on Efficiency

Revise the following sentences to make them

as economical as possible. Count the words in your revised version and compare this number with the word count of the original.

1. Before adding a recreational enterprise, a landowner should determine the nature and extent of the activities contemplated for the enterprise and give some consideration and planning to that enterprise. [29]
2. Robert Koch in 1876 developed a simple test for proving that an organism produced a certain disease, and this test is called Koch's Postulates. [24]
3. The following discussion brings out some considerations which may be helpful in this process. [14]
4. Stockmen should understand that transmission of anaplasmosis can occur when infected blood from one animal is carried mechanically to a susceptible animal. [22]
5. Of course this change in production process will have to be approved in the Naval Production Specification Manual, but to insure NQCD's approval, we will incorporate a 20% increase in testing temperature and in this report will prove that with this increase of temperature, the thermal-testing process will maintain or even improve the degree of quality of the ADI 350 gyro. [62]
6. The purpose of this memo is to supply my recommendation for the operation parameters of the sieve-tray distillation column which will achieve the largest amount of ethanol recovered from the 12% ethanol stream. [34]
7. Since there was no question of the foreman of this crew not knowing the proper procedure, it has been deemed that he was grossly negligent and it is for that reason that he has been laid off since. [39]
8. The witness stated in reply to questions asked during his testimony that the parolee did not say anything while he blocked the door. [22]
9. After receiving the two cylinder heads removed from your race bikes, I began to analyze the failure mechanisms affecting their performance.
10. Attached to this letter you will find a copy of the final project report. [14]

6.5.3 Workshop on Clarity

Revise the following sentences to clarify pronoun references and correct dangling modifiers.

1. Upon attaining maturity, the roughened areas on the bark attract the larvae that pupate in the larger branches.
2. Keeping the soil acid, blue hydrangeas result.
3. Replacing the fuse with a penny, a fire results.
4. Biting the cattle, anaplaxmosis is transmitted by flies.
5. Displaying interest in your guests, your motel rooms will be occupied more often.
6. The doctors base their diagnosis on the nurse's notes; they must be accurate and complete.
7. Nearly three-fourths of the injuries are lacerations and fractures caused by direct contact with the mower's blades; they travel at speeds up to 200 miles per hour.
8. In the brine displacement method energy used in filling the cavern with product can be recovered when it is later removed from the cavern by gravity displacement with brine.
9. Upon arrival officer Brown contacted officer Smith, security officer for the company who informed him that the subject, later identified as Robert Jones, was observed on company property with a gun tucked under his belt. He proceeded to remove

the gun and called the McLean police department for assistance.

10. On September 26, at 7:36 p.m. at the Big Boy Supermarket in Rumsey, subject Ann Taylor was observed by an employee of the supermarket as she placed four packages of meat in her purse which were valued at a total of $11.16 and attempted to leave the store without paying for these items.

At the beginning of this chapter we mentioned that many people in industry unfortunately make themselves stand out because of what they can't do rather than because of what they can do. They can't punctuate correctly, they can't spell, and they can't edit sentences effectively, so they draw attention to themselves. Yet, as we have tried to show in this chapter, the ability to write grammatically correct, direct, efficient, and clear prose is not really very difficult to develop. The patterns of sentences are easy to learn; the conventions of punctuation are not, after all, very difficult; and editing for directness, efficiency, and clarity is not excessively time-consuming. If you are capable of mastering a complex technology, as every competent technician obviously is, you are certainly capable of learning the skills necessary to write effective prose. All that is required is the commitment to work at it carefully and conscientiously over a period of time. Because your reputation as a technician depends to a considerable extent upon your ability as a writer, this commitment is worth your making.

A. Answers for Grammar Review Workshop

6.3.1 Practice with Different Sentence Types

1. When <u>instructors</u> <u>submit</u> the Opscan sheets late or with errors, the <u>system</u> <u>fails</u> to operate. [Complex sentence with the dependent clause first]
2. If the <u>landowner</u> <u>is considering</u> a camp ground, he <u>will find</u> that some campers prefer a primitive site, but <u>he</u> <u>should</u> also <u>realize</u> that most prefer modern conveniences such as flush toilets, electric lights, and appliance hookups. [Compound-complex sentence, dependent clause first]
3. <u>This</u>, however, <u>requires</u> additional time. [Simple sentence]
4. <u>Problems</u> <u>will have</u> to be solved if the cumulative grading <u>report</u> <u>is</u> to be implemented. [Complex sentence, dependent clause last]
5. The problem-oriented health <u>record</u> <u>is</u> a method of documenting health care to promote continuity of patient care among the various members of the health team. [Simple sentence]
6. All <u>notes</u> <u>are signed</u> with each individual's specific credentials, followed by "Respiratory Therapy Department." [Simple sentence]
7. When the <u>bend</u> <u>is made</u> in the plane perpendicular to that of the leads, [<u>you</u> is the understood subject] <u>make</u> the bend at least ⅛ inch from the plastic case. [Complex sentence, dependent clause first]
8. After interviewing those persons directly associated with Community Services, <u>we</u> <u>have determined</u> that the actual cause of failure of the system is the gathering of student information. [Simple sentence]
9. Our <u>study</u> to review the adequacy of the present system <u>began</u> in May 1972 and has just recently <u>been completed.</u> [Simple sentence]
10. A <u>system</u> <u>can function</u> in the manager's absence, but each <u>trainee</u> <u>must be</u> carefully <u>supervised</u>. [Compound sentence]

6.3.2 Workshop on Sentence Punctuation

1. The boy who had been committed for larceny, assault, and truancy was not very cooperative.
2. If a trust has been established, then the corporate trustee provides the level of management competency as well as the continuity.
3. Some grains such as oats, barley, wheat, or sorghum grains may be fed, but they generally present the problem of an inadequate protein intake.
4. After the chicks are six weeks old, heat seldom is required.
5. The doctor bases his diagnosis on the nurse's notes; therefore, these must be accurate and complete.
6. Many nurses are under the impression that nurses' notes are not admissible legal evidence; this is not true.
7. The circuit that is connected to the demonstration board should be tested first.
8. External parasites most likely to attack the broilers are blue bugs, lice, mites, and fleas.
9. Chemical, biological, and cultural controls will be studied.
10. His reaction, crying, was most unusual for an adolescent.
11. That patient, already anxious and depressed, will not react well to questions put in a sharp tone of voice.
12. The problem list now becomes the first page of the record, an index or table of contents to the record.
13. Raising rabbits pays off well, especially

when proper sanitary precautions are taken.

14. The ward was committed for the following offenses: larceny, assault and battery, and truancy.

B.   Answers for Practice in Editing Sentences
The following are correct or reasonable answers; in some cases, other answers are possible.

6.5.1 Workshop on Directness
1. The disturbed youth acted strangely, crawling or huddling on the floor, crying, and chewing on his comb.
2. The defendant's father died of emphysema in 1971, at age fifty-six.
3. The records at Jones High School revealed that the defendant received four Es, one C, and one B during his final semester.
4. In this case, the rigging crew clearly did not understand the use of wire rope clips.
5. The new fuel pumps will be used in test-stand and 50,000-mile endurance testing.
6. This manual method and the mass flow values it provides are causing unreliable test results and extra costs for the research center.
7. Pinhole flaws in the glass lining cause these patchings.
8. The Atomic Energy Commission has requested additional information on the application for a reactor construction permit.
9. Therefore, Midwestern Chemical will be able to save more than $107,000 per year by using the sieve-tray distillation tower as explained.
10. The diesel's lower power-to-weight ratio may also partly account for the diesel's

better fuel economy than the gasoline engine's.

6.5.2 Workshop on Efficiency
1. Before adding a recreational enterprise, a landowner should determine and plan the nature and the extent of the activities contemplated. [20]
2. In 1876, Robet Koch developed a simple test called Koch's Postulate for proving that an organism produced a certain disease. [20]
3. The following discussion identifies considerations which may be helpful in this process. [12]
4. Stockmen should understand that anaplasmosis may be transmitted by blood carried mechanically from an infected animal to a susceptible animal. [20]
5. Of course this change in production process will require approval in the Naval Production Specification Manual, but to insure that approval we will incorporate a 20% increase in testing temperature. This report proves that with this increase the thermal-testing process will maintain or improve the quality of the ADI 350 gyro. [52, in two sentences]
6. This memo recommends operation parameters for the sieve-tray distillation column. These parameters will assure maximum recovery of ethanol from the 12% ethanol stream. [25, in two sentences]
7. Since the crew foreman knew the proper procedure, he was deemed grossly negligent; for that reason he has been laid off. [21]
8. The witness testified that the parolee did not say anything while he blocked the door. [15]
9. After receiving the two cylinder heads from your race bikes, I analyzed the

causes of their failure. [17]
10. The final project report is attached. [6]

6.5.3 Workshop on Clarity
1. Upon attaining maturity, the larvae migrate to the larger branches and pupate in roughened areas on the bark.
2. Blue hydrangeas are produced by acid soil. [or] Blue hydrangeas grow in acid soil. [or] Acid soil produces blue hydrangeas.
3. A fire may result if a fuse is replaced with a penny. [or] If you replace a fuse with a penny, you may cause a fire.
4. Anaplasmosis is transmitted by flies biting the cattle.
5. If you display interest in your guests, your motel rooms will be occupied more often.
6. Because the doctors base their diagnosis on nurses' notes, these notes must be accurate and complete.
7. Nearly three-fourths of the injuries are lacerations and fractures caused by direct contact with the mower's blades, which travel at speeds up to 200 miles per hour.
8. In the brine displacement method, energy used in filling the caverns with product can be recovered when the product is later removed by gravity displacement with brine.
9. Upon arrival, Officer Brown contacted Officer Smith, security officer for the company, who informed him that the subject, later identified as Robert Jones, was observed on company property with a gun tucked in his belt. Officer Smith had removed the gun and called the McLean police for assistance.
10. On September 26 at 7:36 p.m. at the Big Boy Supermarket in Rumsey, an employee saw Ann Taylor place four packages of meat valued at $11.16 in her purse and attempt to leave the store without paying.

C. Correct the following sentences, which all have problems discussed in this chapter.

1. Jim was committed for the following offenses; larceny, assault and battery and truancy.
2. Changes made by the requestor and persons on the approval list shall be reviewed: make any changes necessary to the draft.
3. Typically, the spread of Ich through fingerling ponds is rapid: the young fish usually have gained no immunity and the smaller sizes are less able to produce an immune response.
4. Infected needles for bleeding or inoculating dehorning saw nose tongs tattooing instruments ear notchers or knives for castration all may spread the disease.
5. The patient in Room 39D a diabetic needs special care.
6. The money will be paid in equal installments in October, 1980 and January, April and July, 1981.
7. Mr. Yates refuses to participate in most of the O.T. activities made available to him which he says are due to poor eyesight; therefore it will be necessary to provide Mr. Yates with activities that involve gross motor movement with little demand on visual accuracy.
8. Synthetic bristles usually are made of nylon fiber and are used to apply latex alkyd or oil-based paints.
9. Good initial control can be obtained using only the petal fall and shucksplit sprays however it is usually best to make the pink bud application also since inclement weather may delay petal fall or shucksplit applications.

10. After being handcuffed, the officer stated the parolee calmed down and wasn't abusive.

11. Three consecutive trial runs were made with each aircraft. The first trial was made with full power applied before the brakes were released. The second and third trials without the use of brakes and a gradual application of power from idle to full throttle in five and ten seconds respectively.

12. Our evaluation of Dr. Shepler's investigative procedures lead us to believe that his methods for the surveillance and control of Rocky Mountain spotted fever are effective and thorough.

13. The problem with the current practice is that once in place, it is difficult to seal the edges of the mylar by heat treatment without melting the mylar in the vicinity of the seam.

14. In reviewing the installation procedure, I can see two reasons to alter them.

15. Instead of building the side wall to support water pressure, the construction cost can be lowered by engraving the two ends of the dam into the banks of the stream.

16. In addition the inertness of platinum makes it an ideal choice for this application.

17. Per project guidelines this is being done concurrently with the programming effort (by the particular programmer), I suggested that these people proofread one anothers documents, since they could find most inaccuracies.

18. The only option I see is to make Version A5.1. available to the users here, and possibly having some students test it.

19. The question was raised by Mr. Warren as to whether the on-line requirements of the scheduling program will be compatible with the 64k-byte capacity of the IBM 5110 it is to be run on.

20. In your letter of August 15, you requested my help in determining whether or not Bay Area Bakery can meet the projected increase in market by minimizing existing transportation costs using the current plant facilities to the extent that there would be no need to build a new plant at San Jose.

21. The remote location of the pipeline has little effect on human life and all precautions are being made to preserve an unaltered form of wildlife and landscape.

22. There are a few situations in which heat detectors would be more appropriate.

23. Using a densitometer, measurements would be made on a continuous time basis.

24. This requirement can be met through an inexpensive fire detector in each home, however this is an oversimplification of our problem.

25. He took the parolee home and was invited inside and then he fell asleep on a couch and, then, shots were fired.

7.1 Introduction

Have you ever been sick and asked your roommate to record your American history class on your cassette recorder? The next day, you settle back in the chair, turn on the cassette, and plan to do college the easy way. *Hiss, whistle, sniffle.* As you play the tape, you get more of the class than the professor—the February cough, crumpling notebook pages, wisecracks about the lecture, whispered discussions about Saturday night. Unlike you, the recorder is unselective. The tape gives you more of America today than America of yesterday. In electronics jargon, you get as much noise as signal.

Poorly edited paragraphs and sentences are like noise-filled passages on a tape. They interfere with your message and irritate your readers. First-draft paragraphs invariably are full of all kinds of noise, because actual writing is done at the paragraph level. The paragraph therefore is where editing begins. You edit to filter out the noise and tune in the signal.

You should actually do paragraph editing and sentence editing at the same time. Shape the structure of each sentence with the design of the entire paragraph in mind. A variety of sentence structures can help express any idea. The appropriate sentence structure is determined partly by the content of the sentence and also by the context—how the idea in that sentence relates to the ideas in the other sentences in the paragraph. During paragraph editing, you clarify the context and edit single sentences.

This chapter is concerned with paragraph design. Of course, you should apply these principles when you write your first-draft paragraphs, so that you do not introduce too much noise into the system to start with.

However, apply these principles systematically when editing, in order to transform the first-draft scrawls into final-draft prose ready for the typist.

This chapter explains how to:
1. Establish the main point of a paragraph by writing a core sentence.
2. Establish general-to-particular order in a paragraph.
3. Establish a pattern of individual sentence structures in a paragraph.
4. Edit each paragraph to establish an appropriate paragraph pattern.
5. Use special paragraphs and formats.

The first four of these skills will help you prepare your first draft; the last helps you prepare your final draft. When you master these skills, your final and even your first-draft paragraphs will be improved considerably.

7.2 The Single-Idea Paragraph

When sentences work together to develop the aspects of a single idea, they form a unified whole called a paragraph. In technical reports, the paragraph is a sequence of sentences that presents and explains one idea. In appearance, the paragraph is a block of prose; paragraphs are the building blocks of a report. In this chapter, we discuss this block as a distinct entity—its appearance, its components, and its intellectual shape.

The single idea of the paragraph is presented in one sentence, the *core sentence.* It comes at the beginning of the paragraph and therefore establishes the general-to-particular order of sentences in the paragraph. The core sentence and the paragraph are brought to the reader's attention by white space—vertical white space between paragraphs in single-spaced format, sometimes accompanied by horizontal white space created by indenting the first line of the paragraph.

7.2.1 The Core Sentence

Each paragraph in a technical report should have a core sentence, which explicitly states the main idea explained and developed by the other sentences in the paragraph. The core sentence is the most general statement in the paragraph; the other sentences amplify, develop, and explain in detail the meaning of the core sentence.

Here is an example of a technical report paragraph, with the core sentence in *italics:*

> *To reduce the probability of mercury poisoning in the labs, additional safety measures can be taken.* First of all, smoking, drinking, and eating should not be allowed in labs where mercury is handled regularly. Smoking is particularly dangerous because mercury vapor is inhaled directly. Second, an instrument to measure the amount of mercury vapor present should be purchased. The vapor should be measured during operation and after cleanup. Third, a special exhaust system should be installed to control the mercury vapor at the sources. The system should have a filtering system to collect the mercury before exhaust is released to the outside environment.

The main clause of the core sentence in this paragraph states the main idea: "additional safety measures can be taken." This is the most general statement in the paragraph. All the other sentences explain the main idea in detail.

The paragraph in a technical report is often treated differently from the paragraph in other types of writing, such as the personal-experience essay. In the personal essay, a paragraph often explains and develops an idea indirectly. That is, the essay paragraph can convey an idea without ever explicitly stating it. For example, you could explain the feeling of your first door-to-door encounters while doing political work without stating that feeling directly, because the direct statement might prevent the reader from experiencing what you are trying to convey. Such an approach is inappropriate for technical reports.

The core sentence comes at the beginning, so that your reader is never in doubt about the main point of the paragraph. This allows the reader to scan the paragraph quickly or skip the particulars altogether.

The paragraph on safety precautions to prevent mercury poisoning illustrates the general-to-particular order. The clause "additional safety measures can be taken" establishes an explanatory pattern. Three sentences explain in detail what the writer means by "additional safety measures." Notice that each of the three is followed by an even more particular sentence, explaining and amplifying the point. In this paragraph, the core idea is supported by three particulars, each of which in turn rests on other particulars (see Table 7-1).

7.2.2 Editing to Provide a Core Sentence

Your first editing task is to examine each paragraph in your discussion to make sure it has a core sentence and to put that core sentence at the beginning. You wrote the first draft while you were still formulating your ideas, so do not be surprised that this first editing task is required. Some of your paragraphs will lack core sentences; others will have core sentences, but in the middle or at the end. More of your first-draft paragraphs will have core sentences at the beginning, as you become more experienced as a writer. For now, however, examine each paragraph and revise, reposition, or write the core sentence.

Examine each paragraph to identify the sentence with the highest level of generalization. That will become your core sentence. (If you are unsure about generalization, review Chapter 4.)

If you do not find a generalization that sums up all the ideas in the paragraph, you must write a core sentence. Sometimes first-draft paragraphs are entirely particulars:

The fuel system did not leak at impact during the 30-minute observation period. The fuel system did not leak during the pressure check. The oil

Table 7-1
Paragraph Seen as a Core Sentence Supported by Particulars

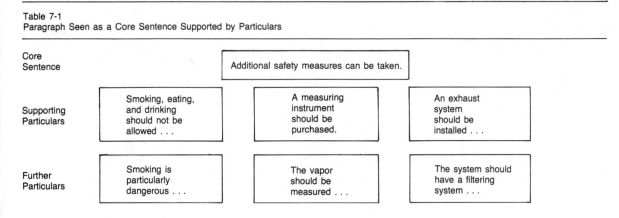

Core Sentence		Additional safety measures can be taken.	
Supporting Particulars	Smoking, eating, and drinking should not be allowed . . .	A measuring instrument should be purchased.	An exhaust system should be installed . . .
Further Particulars	Smoking is particularly dangerous . . .	The vapor should be measured . . .	The system should have a filtering system . . .

consumption valve was in close contact with the fuel return line, but did not damage it.

This paragraph is all particulars, with no core sentence. The core sentence of this report was added during revision, permitting other changes to be made in order to tie the whole paragraph together:

> The fuel system performed satisfactorily in impact testing. It did not leak at impact during the 30-minute observation period, and it did not leak during the subsequent pressure check. The fuel return line, although in close contact with the oil consumption valve, was not damaged by the impact.

Putting a core sentence first is a simple matter of rearranging sentences. Usually all that is required is a slight revision to reconnect the sentences with new transitional words and phrases. The following paragraph can easily be rearranged to put the core sentence first. In the first-draft version, the core sentence came last:

> Burrowing bugs migrate into peanut fields around midsummer. They attack the nuts developing beneath the soil surface. They feed only on young and maturing peanuts. Their feeding causes a dark brown mottling of the kernels and results in grade reductions. *The burrowing bug thus is economically damaging to peanut crops.*

Because the core sentence appears at the end, the reader does not know what the paragraph is really about until after the third sentence. Up to that point, the paragraph is only about how the burrowing bug feeds. When the core sentence comes first, the reader is not misled:

> *The burrowing bug is an economic problem because it reduces the value of the peanut crop.* Burrowing bugs migrate into peanut fields around midsummer. They attack the nuts developing beneath the soil surface, and they feed only on young and maturing nuts. Their feeding causes a dark brown mottling of the kernels, which results in grade reductions. The grade reduction is the primary economic loss.

Now, the feeding habits of the burrowing bug are put into context. The paragraph becomes a cause/effect paragraph, explaining the core sentence. Notice how the new first and final sentences clarify and restate the core idea.

7.3 The Pattern of the Paragraph

The paragraph is a unified group of sentences developing one core idea. The paragraph is unified in another sense as well: its sentences form a pattern.

A paragraph pattern has two characteristics. First, the ideas are presented in a logical order. Second, this logical order is made clear by the structure of the sentences. These characteristics can be seen when you reflect on how you write your sentences. You put them down in paragraph form by writing them one after another. Each sentence states a point, and you move from one point to another according to the logic of your explanation or argument. You don't write randomly. The sequence of the sentences and their subject-verb relationships reflect the logic of the argument or explanation. Your subjects and verbs make clear how one sentence relates to the others and how the whole paragraph fits together.

7.3.1 Establishing Pattern Through Sentence Structure

The structures of the sentences in the paragraph have a logical relationship to each other, like circuits on a circuit board. For example, examine this three-sentence paragraph, whose transitions have been eliminated to clarify the pattern:

The fuel system performed satisfactorily.
The fuel system did not leak after impact.
The fuel system did not leak during the pressure check.

The order of the sentences is determined by the logic of the explanation—the general statement, the impact examination, then the pressure check. The pattern is one of parallel subjects and parallel predicates:
Subject 1 + Predicate 1
Subject 1 + Predicate 2
Subject 1 + Predicate 3

The subject of all three sentences is "the fuel system." The first predicate is general; the second and third predicates are more particular. That is, "performed satisfactorily" tells many readers all they need to know about this paragraph, while "did not leak after impact" and "did not leak during the pressure check" provide the specific information some readers have to have. Notice that the second and third predicates are on the same level of generality; in this case, their order is determined logically by the fact that the impact test was performed before the pressure check.

If the paragraph had a logical order but no pattern, the meaning would not have been presented quite so directly:

The fuel system performed satisfactorily.
Leakage was not observed after impact.
The pressure check did not cause a leak.

In this set of three sentences, the subjects and verbs do not form a pattern. The subject of the first sentence, the main topic of the paragraph, is not mentioned again. With the writer's shift into passive voice in the second sentence, the

predicate of the first sentence "did not leak" becomes the noun subject of the second ("leakage"). The real subject and predicate almost disappear from the third sentence, with only the word "leak" preserved, in noun form. The third sentence shows complete lack of pattern; neither the subject nor the verb is the focus of the paragraph. Paragraphs which lack a pattern, such as this example, force the readers to make connections, to read between the lines, and to add thoughts of their own. Without a pattern, the ideas in a paragraph get out of focus.

7.3.2 Establishing Pattern Through Editing

To establish a pattern, you first edit the sentences to focus on the main idea, then you order the sentences logically, editing them one by one to form a pattern. This requires you to look at the subjects and verbs, eliminating indirect constructions. Finally, you introduce stylistic and transitional elements to make the sentences fit together smoothly.

Your first task is to make certain that every sentence focuses on the main idea. Use the core sentence to control the other sentences. You examine each sentence to make sure it speaks directly to the main point of the paragraph. If a sentence isn't really about the main point, leave it out; if it doesn't quite address the main point, revise it.

Notice that the final sentence in the following paragraph is not really about the main point and that the fifth sentence is not focused clearly:

Because mercury spills pose the most serious safety hazard, provisions must be made for cleaning up mercury spills in the laboratory. To contain the spills and simplify cleaning operations, each lab bench where mercury is used should be enclosed. Mercury often splashes and spills while being poured, because of its great density, surface tension, and low viscosity. Droplets of mercury will scatter in all directions and roll into cracks and corners. A sheet-linoleum floor covering with sealed joints and coved edges can be used. The spilled mercury can be recovered by a suction apparatus. Employees should take precautions to avoid unnecessary mercury spills.

The final sentence on handling belongs in a paragraph about employee training, not in this one, which is about what to do when spills occur. The fifth sentence needs to be revised so that it directly addresses the main point of the paragraph, how to clean up spills. It could read, "A sheet-linoleum floor covering with sealed joints and coved edges would eliminate cracks, which trap mercury."

Once you have established the focus, form the paragraph pattern. The structure of the core sentence establishes a logical connection between the ideas it introduces. For example, the core sentence in the following paragraph establishes a problem/solution sequence. Mercury spills are the problem; provisions to clean up mercury spills are the solution. The sentences in the paragraph should be rearranged in that logical sequence:

(1) *Because mercury spills pose the most serious safety hazard, provisions must be made for cleaning up mercury spills in the laboratory.* (2) Mercury often splashes and spills while being poured, because of its great density, surface tension, and low viscosity. (3) Droplets of mercury form and scatter in all directions and roll into cracks and corners. (4) To contain the spills and simplify cleaning operations, each lab bench where mercury is used should be enclosed. (5) A sheet-linoleum floor covering with sealed joints and coved edges can be used to eliminate cracks, which trap mercury. (6) The spilled mercury can then be recovered by a suction apparatus.

The core sentence comes first. The second and third sentences go into the problem in more detail, and the fourth through sixth sentences present the solution. To rearrange the sentences in logical sequence, the second sentence of the original paragraph was moved, becoming the fourth sentence of the new paragraph. The core sentence implies that the writer will talk first about mercury spills, then about means of coping with the spills. The solution seems logically ordered, from containment to convenience to actual cleanup.

To edit each sentence so that it reflects the logical sequence, examine the sentences in the paragraph. The subjects should not be chosen randomly, but instead should be related. The same is true of the predicates; one should be related to the others. If the paragraph focuses on a fuel system, then perhaps "fuel system" or a suitable pronoun should be the subject of every sentence. If the paragraph focuses on the performance of the fuel system, then "did not leak under pressure" and similar verbs, objects, and modifiers should be the predicates.

Mentally put the complete subjects, verbs, and complements of these sentences in columns and examine the relationships of the parts of these sentences. Mentally putting the sentences in the paragraph on mercury spills, for example, in columns, as in the table below, helps you to see where the editing needs to be done:

This indicates that the subjects of the first three sentences clearly develop the concept of "mercury spills." However, there is a problem in the fourth through sixth sentences, whose subjects change for no reason. These sentences need to be edited. The last three sentences present the solution, referring to the second clause of the core sentence, "provisions must be made for cleaning up mercury spills in the laboratory." This clause should control those sentence structures (see table below).

The fourth sentence is the same as it was, except that the infinitive phrase ,"to contain the spills and simplify cleaning operations" has been placed at the end, to keep the sentence parallel to the second clause of the core sentence. The fifth sentence was already parallel, so it has been left alone. The last sentence has been changed from passive to active. So the three solution sentences parallel the solution clause of the core sentence, just as the second and third sentences parallel the problem clause of the core sentence. This results in a clear paragraph pattern:

Because mercury spills pose the most serious safety hazard, provisions must be made for cleaning up mercury spills in the laboratory. Mercury often splashes and spills while being poured, because of its great density, surface tension, and low viscosity. Droplets of mercury form and scatter in all directions and roll into

Subject	Verb	Object or Modifier
(1) . . . mercury <u>spills</u> <u>provisions</u>	<u>pose</u> <u>must be made</u>	the most serious safety hazard, for cleaning up mercury spills in the laboratory.
(2) <u>Mercury</u>	. . . <u>splashes</u> . . . <u>spills</u>	while being poured. . . .
(3) <u>Droplets</u> of mercury	<u>form</u> . . . <u>scatter</u> . . . <u>roll</u>	in all directions into cracks and corners.

(4)
each lab <u>bench</u> where
    mercury is used
	<u>should be enclosed.</u>	To contain the spills and
		    simplify cleaning operations,

(5) A sheet-linoleum floor <u>covering</u> . . .	<u>can be used</u>	to eliminate cracks, which trap mercury.
(6) The spilled <u>mercury</u>	<u>can . . . be recovered</u>	by a suction apparatus.

Subject	Verb	Object or Modifier
(1) . . . <u>provisions</u>	. . . <u>must be made</u>	. . . for cleaning up mercury spills in the laboratory.

. . .
(4)
Each lab <u>bench</u> where
    mercury is used
	. . .
	<u>should be enclosed</u>
		. . .
		to contain the spills and
		    simplify cleaning operations.

(5) A sheet-linoleum floor <u>covering</u> . . .	<u>can be used</u>	to eliminate cracks, which trap mercury.
(6) A suction <u>apparatus</u>	<u>can . . . recover</u>	the spilled mercury.

cracks and corners. Each lab bench where mercury is used should be enclosed to contain the spills and simplify cleaning operations. A sheet-linoleum floor covering with sealed joints and coved edges can be used to eliminate cracks, which trap mercury. A suction apparatus can then recover the spilled mercury.

Editing techniques that enable you to edit sentences one by one to form a good paragraph pattern are the use of parallelism, the use of true subjects and verbs, and the elimination of indirect constructions.

Parallelism, your most important editing concept, is the repetition of a grammatical structure, along with certain key words. Parallelism makes concepts clear and establishes relationships, as in the three sentences on fuel-system performance:

> The fuel system performed satisfactorily.
> The fuel system did not leak after impact.
> The fuel system did not leak during the pressure check.

Once the pattern is clearly established, pronouns and transitional words provide stylistic variety without breaking the pattern:

> The fuel system performed satisfactorily. It did not leak after impact, and it did not leak during the subsequent pressure check.

As much as possible, you should make the true subject the grammatical subject and the true verb the grammatical verb. In the sentence "The fuel system did not leak after impact," the true subject is the grammatical subject and the true verb is the grammatical verb. Before editing, the sentence was "Leakage was not observed after impact." The grammatical subject, "leakage," was not the true subject, and the grammatical verb, "was not observed," was not the true verb. The core

idea indicates what the true subject and true verb are. Your knowledge of sentence grammar will help you make these the grammatical subject and the grammatical verb.

Concern for parallelism and true subjects and verbs requires you to eliminate indirect constructions. An example of an indirect construction is the "it . . . that" construction, which forces the main idea of a sentence into a subordinate clause. The sentence "It has been concluded that the fuel system performed satisfactorily" has an unnecessary "it . . . that." "It" is not the true subject, and "has been concluded" is not the true verb. The statement "The fuel system performed satisfactorily" is a conclusion that puts the main idea directly to the reader. Avoid such common "it . . . that" constructions as

> "It has been concluded that . . ."
> "It can be seen that . . ."
> "It follows that . . ."
> "It is clear that . . ."

The final step in editing to establish a pattern is using transitions. Transitions, words and phrases that modify the entire sentence, clarify the paragraph pattern for the reader and help the sentences flow smoothly.

The most important function of transitions is explaining the paragraph pattern to the reader. If the paragraph lists additional safety measures, the sentences will start with such transitional words as "first," "second," and "third." If the paragraph explains an event, the sentences will have time-signal transitions, such as "initially," "following the admissions check," and "before the exam can be completed." If the paragraph explains a cause and its effect, the sentences will have causal connectors, such as "because patients are anxious" and "the result of mixing the solutions." Logical relationships of ideas are signalled by such words and phrases as "however" and "in addition." When you edit a paragraph, introduce and use transitional words and phrases appropriately and consistently.

Often, the first-draft transitions will be inappropriate, inconsistent, or incomplete. They can be inappropriate if they fail to signal the exact relationships between ideas. A common example of an inappropriate transition is the use of "and" when the exact relationship of ideas would be better expressed with "but." Common examples of inconsistent transitions are the series "first," "second," . . . "finally" and "first," "secondly," . . . "also." Transitions are incomplete when the paragraph signals a pattern with "first" or "the cause of," for example, but does not follow the pattern with "second" or "the effect of."

Each sentence within the paragraph should seem to flow smoothly into the next. Without transitions, sentences move like freight cars when the locomotive starts to pull them; they jerk into action. Sentences often still have this effect after you have edited them to form the paragraph pattern. To eliminate the jerkiness while still preserving the pattern, add a few transitional words and phrases, change a few words, and move a few phrases. In the following paragraph, the words and phrases in *italics* illustrate some of these techniques.

> Debeaking *after 17 weeks* can cause problems *for pullets*. Debeaking *usually* causes pullets to lose body weight, *which* can be detrimental if the loss occurs during the sexual maturing process. Loss of body weight *during this critical period* has been implicated as a cause of low production peaks. Debeaking *also* causes problems *if it occurs* at housing. Debeaking causes beak tenderness, and pullets *therefore* have difficulty adjusting to waterers *at this time. Failure to adjust*

results in increased mortality, excessive weight loss, smaller eggs, and lower production rates.

Transitional words and phrases allow you to employ repetition of words and phrases, parallel sentence structures, and signals of paragraph pattern without making the paragraph monotonous. Well-chosen transitional words and phrases often perform all these functions simultaneously.

7.3.3 Choosing the Appropriate Pattern

When you edit a paragraph, you should make sure that its sequence is appropriate as well as logical. At the same time that you edit a paragraph to give it a pattern, you make certain this pattern is the appropriate one.

As we have said, the sentences in a paragraph should have a logical sequence, a pattern. As you undoubtedly have noticed, this logical sequence can be of different types. These are generalized patterns which all of us use and recognize, at least unconsciously. To write and edit effective paragraphs, you must consciously recognize and choose patterns which are appropriate to your purposes.

The eight basic types of paragraph pattern are the same as those for the particulars of the discussion:
Persuasive Patterns
    Persuasion
    Problem/Solution
    Cause/Effect or Effect/Cause
    Comparison and Contrast
Informative Patterns
    Analysis
    Description
    Process, Causal Chain, and Instructions
    Investigation

These patterns have similar basic structures:

Persuasive Pattern

General	Core sentence states the main point.
↓	
Particular	Particulars support or document the main point.

Informative Pattern

General	Core sentence states the main point.
↓	
Particular	Particulars explain the main point.

In many paragraphs, these patterns are simple and direct, but some paragraphs combine patterns, so that the basic pattern is made up of several subpatterns.

7.3.3.1 Persuasion

You use the persuasion pattern when your explicit purpose is to support a conclusion. The pattern begins with a direct statement of the conclusion, then presents the primary support for that conclusion. Secondary support, qualified support, or explanation of the primary support follow. Rebuttal of other evidence, a form of qualification, comes last. Quite often, the conclusion is restated if any negative discussion has been introduced. The persuasion paragraph can be outlined as follows:

1. Statement of the conclusion
2. Primary support for the conclusion
3. Secondary support, qualified support, explanation of primary support
4. Rebuttal of other support or negative discussion, if any
5. Restatement of conclusion, if necessary

Here is an example of the persuasion paragraph:

> The Board is satisfied on the evidence that Smith did violate his parole. Sears security guard T. Sawyer testified that she saw John Smith hide a record in his coat and that when she and other employees chased Smith, she saw the record fall from his clothing. This testimony was supported by the testimony of L. Olsen, a Columbus policeman who works as a part-time security guard at the store.

This paragraph has a simple three-sentence pattern: statement of conclusion, primary support, and secondary support.

7.3.3.2 Problem/Solution

You use this pattern to demonstrate that you have solved a problem. The pattern begins with a statement of the problem. If necessary, the problem is explained. Then a point-by-point explanation of the solution is presented, either in decreasing order of importance or in another logical sequence. The problem/solution paragraph can be outlined as follows:

1. Statement of the problem
2. Explanation of the problem, if necessary
3. Point-by-point explanation of the solution

Here is an example of the problem/solution paragraph:

> Because mercury spills pose the most serious safety hazard, provisions must be made for cleaning up mercury spills in the laboratory. Mercury often splashes and spills while being poured, because of its great density, surface tension, and low viscosity. Droplets of mercury form and scatter in all directions and roll into cracks and corners. Therefore, each lab bench where mercury is used should be enclosed to contain the spills and simplify the cleaning operations. A sheet-linoleum floor covering with sealed joints and coved edges should be used to eliminate cracks, which trap mercury droplets. A suction apparatus can then recover the spilled mercury.

This paragraph explains what the problem is, then presents the solution in a logical, point-by-point sequence of sentences.

7.3.3.3 Cause/Effect and Effect/Cause

You use a cause/effect pattern when you wish to "prove" a cause or an effect. You begin with a statement of the cause/effect relationship. This will have one of two forms:

either that A causes B or that B results from A. Then you explain how A causes (or will cause B) or how B results (or will result) from A. This explanation usually follows the logical cause/effect sequence, although sometimes it runs in decreasing order of importance. The cause/effect paragraph can be outlined as follows:

1. Statement of the cause/effect relationship
2. Explanation of the sequence from cause to effect
   (or)
   List of causes or effects in decreasing order of importance
   (or)
   List of causes or effects in decreasing order of probability
3. Statement of the cause or effect to be shown

Here is an example of the cause/effect paragraph:

> The burrowing bug is an economic problem because it reduces the value of the peanut crop. Burrowing bugs migrate into peanut fields around midsummer. They attack the nuts developing beneath the soil surface, and they feed only on young and maturing nuts. Their feeding causes a dark brown mottling of the kernels, which results in grade reductions. The grade reduction is the primary economic loss.

Notice that, whether you wish to explain a cause or an effect, the sequence of particulars goes from cause to effect. The exception to this is a list of either causes or effects in decreasing order of importance.

7.3.3.4 Comparison and Contrast
You most often use this pattern in order to show reasons for a choice among alternatives. At times, you also use it to clarify the reader's understanding of something by detailing its similarities to and differences from another thing. Begin by stating the conclusion; then present a point-by-point comparison and contrast in decreasing order of importance. The comparison and contrast paragraph has the following outline:

1. Statement of the conclusions of the comparison and contrast
2. Point-by-point comparison or contrast in decreasing order of importance
3. If necessary, point-by-point contrast or comparison, whichever has not already been done

Here is an example of a comparison-and-contrast paragraph:

> On the basis of these findings, I recommend that this department use Mini Wear Type I carbon paper rather than Type II carbon paper. The legibility and the long-lasting and noncurling qualities of Type I surpass those of Type II. Because it is longer lasting, Type I carbon paper also will be more cost-efficient.

This paragraph ends with a synthesis or interpretation of the contrast, a variation of the restatement of conclusion technique used within any paragraph pattern.

7.3.3.5 Analysis
You use this in order to explain a thing or concept by breaking it down into its components. Begin by stating the whole as the sum of its particulars. Then explain the components one by one, either in decreasing order of importance or according to a logical relationship among the particulars. The analysis paragraph can be outlined as follows:
1. Statement of the whole
2. Point-by-point explanation of the parts or components, either in decreasing order of importance or according to their logical relationships

Here is an example of an analysis paragraph:

To reduce the probability of mercury poisoning in the labs, additional safety measures can be taken. First, smoking, drinking, and eating should not be allowed in labs where mercury is handled regularly. Smoking is particularly dangerous because mercury vapor is inhaled directly. Second, an instrument to measure the amount of mercury vapor present should be purchased. The vapor should be measured during operation and after cleanup. Third, a special exhaust system should be installed to control the mercury vapor at the sources. The system should have a filtering system to collect the mercury before exhaust to the outside environment.

In this paragraph, the concept of additional safety measures is broken down into three particular ideas. Smoking, eating, and drinking are mentioned first either because they are most important or perhaps because they are the most immediate means of controlling the problem. The second and third particulars have a logical relationship to each other.

7.3.3.6 Description

You use this pattern when you want to explain how some physical device works. Begin with an overview of the object, then present a part-by-part description of the object, arranged according to the functional relationships among these parts. The description paragraph has the pattern:

1. Overview of the object
2. Part-by-part description according to the functional relationship among the parts

Here is an example of a description paragraph:

For chemofishing sampling, a blocking net encloses a measured area so that all the game fish can be trapped and killed with rotenone. The net is 150 meters long and 8 meters deep, and it is placed where the depth, vegetation, and current will not impair efficiency of the sampling. Bamboo stakes at the corners of the net hold it in place. First, the ends of the net are staked at the shoreline, from 25 to 50 meters apart. Then the net is staked in the reservoir to form a rectangular enclosure 1250 meters square. The float line of the net has 12.5-cm oval cork floats 50 cm apart, and the lead line has .05 kg lead sinkers 75 cm apart. The lead line must be checked by scuba gear to make sure it rests on the bottom. The webbing of the net has a 2.54-cm square mesh. The blocking net therefore traps all juvenile and adult fish, which are killed with rotenone. The standing crop in kg/ha then is calculated.

Notice how purpose or function determines the selection of particulars in a description paragraph. Function also controls the spatial arrangement of the particulars.

7.3.3.7 Process, Causal Chain, or Instruction

You use such a paragraph when you want to explain a sequence of events in time. The process paragraph is used primarily to explain how to do something, although it can also be used, like description, to explain how an object works. Begin by stating the purpose of the process. You also may have to present an overview of the process, by identifying its major stages. Then present a step-by-step explanation of the process, making the purpose of each stage clear. The process paragraph has the following outline:

1. Statement of the purpose of the process
2. Overview of the major stages of the process, if necessary
3. Step-by-step explanation of the process

Here is an example of a process paragraph:

To look their best in exhibition, birds must be washed in warm water with a mild detergent, then thoroughly rinsed and dried. Wash a bird patiently, rubbing suds with, not against, the feathers. Care must be taken to remove all grease and dirt without ruffling the feathers. Then rinse the bird two or three times with clean, warm water. Take care to remove all the detergent, or else the feathers will stick together when they dry.

A small amount of bluing in the last rinse will bring out the whiteness of the feathers, but too much will turn them blue. Dry the bird thoroughly with a soft towel, again taking care not to ruffle the feathers. Place the bird in a clean, dry coop in a warm room. Birds must be washed at least twenty-four hours before showing.

The first sentence in this paragraph states the method of the process and presents an overview of its major stages. The step-by-step explanation is clearly presented in functional contexts.

7.3.3.8 Investigation

Use this paragraph to explain how you conducted an investigation. The investigation paragraph, a variation of the process paragraph, deserves separate attention. This type of paragraph often stands by itself in a technical report under the heading "Test Procedure." Begin with a statement of the purpose and general method of the investigation, then present the details in the step-by-step sequence of the experimental method. The investigation paragraph has the following outline:

1. Statement of the purpose and general method of the investigation
2. Itemization of materials, methods, or specifications, if necessary
3. Presentation of results
4. Analysis of results, if necessary
5. Formulation of conclusions, if appropriate

Here are two examples of investigation paragraphs from the same report:

I performed density tests both on a one-foot layer of dense, brown, clayey sand and on a dense, uniform, fine sand found beneath the surface layer at the Delhi Road site. Both layers had relative densities of close to 100%. Since constuction standards consider a relative density of 70% adequate for vibratory loads, very little settlement would occur at the Delhi Road site.

The *in-situ* density was determined by the "sand cone" and "drive sampler" methods. These were the standard ASTM tests D1556 and D2937 respectively. For the fine sand, the following results were computed:
Dry density: 102 #/ft.^{3}
In-situ void ratio: 0.61

Notice that the investigation paragraph presents only those stages of the experimental method appropriate for the specific purpose of the paragraph. All of the stages are usually presented in the investigation segment (Chapter 5).

7.3.4 Pattern and Subpattern

In the previous section, we presented separate outlines and examples for eight basic paragraph patterns. In actual practice, of course, many paragraphs combine two basic patterns, subordinating one of them as a subpattern of the other. The combinations of patterns are so numerous there is no point in attempting to list them; nonetheless, it may be useful to illustrate how easy it is to combine patterns. Once you see how they can be combined, you can use the eight basic paragraph patterns to give a paragraph an appropriate pattern and subpattern. Here is a paragraph with a pattern and a subpattern:

Debeaking after seventeen weeks can cause problems for pullets. Debeaking usually causes pullets to lose body weight, which can be detrimental if the loss occurs during the sexual maturing process. Loss of body weight during this critical period has been implicated as a cause of low production peaks. Debeaking also causes problems if it occurs at housing. Debeaking causes beak tenderness, and pullets therefore have difficulty adjusting to waterers at this time. Failure to adjust results in increased mortality, excessive weight loss, smaller eggs, and low production rates.

This paragraph's basic pattern is analysis. It itemizes two problems caused by debeaking:
Debeaking causes problems:
    Problem 1: Debeaking causes weight loss.
    Problem 2: Debeaking causes beak
      tenderness.

Each problem, however, has a cause/effect subpattern:
    Problem 1 causes low production peaks:
      weight loss $\longrightarrow$ sexual maturation
      complications $\longrightarrow$ low production peaks.
    Problem 2 causes increased mortality, etc.:
      beak tenderness $\longrightarrow$ difficulty adjusting to
      waterers $\longrightarrow$ increased mortality, etc.

Once you understand the structure of the patterns, you should have no difficulty in devising an appropriate pattern and subpattern for any paragraph.

7.3.5 Special Paragraphs

No doubt you have realized that our discussion of core sentences and paragraph patterns does not cover quite all the paragraphs used in technical reports, for example, transition paragraphs, purpose-statement paragraphs, conclusions-and-recommendations paragraphs, and results paragraphs. All of these differ from the eight basic patterns we have discussed. However, these types of paragraphs also can be formatted in outline form.

A transition paragraph shows the connection between paragraphs and between parts of a report. It signals the end of the part that has gone before and shows the relationship of what is to come to what went before. Transition paragraphs usually lack core sentences. Here is an example of a transition paragraph:

> Proper housing is a prerequisite for successful rabbit raising. Good equipment is necessary, as

well. Not much equipment is required, but feeding, water, and nesting equipment must be adequate and sanitary.

This paragraph provides a transition between the paragraphs on housing and those on equipment. Although transition paragraphs do not contribute much meaning to a report, they do clarify its structure and thus make it easier to read and understand. Transition paragraphs often accompany formatting devices which also clarify the structure of the report, such as white space, numbering, and headings (Chapter 8).

Other special types of paragraphs also are common in technical reports. In Chapter 3, we discussed purpose statements; in Chapter 4, we discussed conclusions and recommendations. Although the communication purpose statement provides the core idea for a whole report, the paragraph itself, as you have seen, lacks an internal core sentence. Similarly, the conclusions and recommendations paragraphs often contain no core sentence. Nevertheless, the purpose statement paragraph usually follows the process sequence, and the conclusions and recommendations paragraph usually follows the investigation sequence. Thus, even without core sentences, these paragraphs have recognizable patterns.

Special types of paragraphs which we have not discussed in particular but which are common in technical reports are paragraphs with headings such as "Test Equiqment," "Specifications," or "Test Results." These paragraphs usually have no core sentences, but they clearly follow a pattern, such as an analysis pattern. Here is an example of such a paragraph:

> *Criteria:*
>   Three criteria are used to evaluate the results:

1. Cost of energy. Our data were formulated using average market prices for the quarter ending 31 March 1977. However, the cost of purchase of a unit of energy will increase with time.
2. Outside temperature. Our investigation used the outside temperature range of $-20°F$ to $70°F$. According to the Jackson County Weather Bureau, this is the range over the academic year.
3. Temperature control. Our calculations show that $70°F$ for daytime operation and $65°F$ for nighttime optimizes efficiency of energy use.

You probably have noticed that headings and other format devices often replace core sentences in special types of paragraphs. As in the example above, in fact, these special "paragraphs" sometimes appear in the form of a list of specifications. Such a "paragraph" may even be a table of results that is not explained in complete sentences. Here is an example of such a format:

RESULTS OF THE INVESTIGATION

	No. of Times Carbon Paper Used	No. of Legible Copies Produced	Erasability	Curling
Mini Wear Type I (Hard Carbon)	20	10	Difficult	None
Mini Wear Type II (Soft Carbon)	15	8	Average	Slight

In short reports, conclusions and recommendations as well as cost-benefit figures also are often presented in tabular or outline form.

Special paragraph formats are but an extension of paragraph editing in technical reports. Your primary concern is to present information clearly and efficiently, which requires order, consistency, and conciseness.

Exercises

A.   Using the procedure you have been given for editing paragraphs, evaluate the following paragraphs and suggest improvements. Specifically, do the following for each paragraph:

1. Identify the core sentence and suggest any necessary revision to make the core sentence directly state the main point of the paragraph. If you cannot identify a core sentence, formulate one.

2. Identify the paragraph pattern (and subpattern, if any) and determine whether the pattern is the one suggested by the core sentence. If you think it is not, suggest an appropriate pattern.

B.   Edit three of these paragraphs or others assigned to you to establish patterns. Use your revised core sentences and the appropriate patterns from Exercise A:

1. Establish focus by making sure every sentence speaks directly to the main point of the paragraph.

2. Put the sentences in a logical sequence.

3. Edit the sentences by the technique of putting the subjects, verbs, and objects or modifiers in columns.

4. Introduce and revise appropriate transitions.

a. Lesser peach-tree borers are seldom a problem in well-tended orchards. They are attracted to damaged areas in the tree, which usually result from poor cultural practices. Careless pruning provides many favorable sites for egg-laying around the stubs of branches. Shape trees properly while they are young, leaving well-spaced, open crotches and eliminating the need for making large cuts later. Branches broken because of insufficient thinning of the fruit and wounds arising from barking the tree with equipment during cultivation and harvesting also are attractive to the lesser peach-tree borer. Prevent sun-scalded areas, which sometimes afford entry for the larvae, by leaving small branches to shade large limbs. Winter injury of trunk and scaffold branches and cracked limbs, resulting from scale infestation, also provide sites for larval feeding.

b. Although a corporation is a legal entity, it does not, however, have all of the rights, privileges, and obligations of an individual. For example, it cannot vote except in some elections such as an irrigation district, where it votes as a member of the district rather than as an individual. Even though a corporation cannot be imprisoned, this does not protect its owners and officers from such treatment. A limiting feature of a corporation, particularly as a farming venture, is that it cannot go freely from state to state doing business in each state on the same terms as the business in that state. A corporation must be registered in each state in which it does business, and it must operate within the laws of each state regulating such business.

c. Economic growth of a farming or ranching business may be enhanced by combining assets of two or more individuals in a partnership. However, the stability of the firm may be in question because of the possibility of willful termination by a partner or involuntary termination caused by the death or retirement of a partner. Involuntary termination may be guarded against through specific provisions in the partnership agreement which allow the partnership to continue with the estate as a partner or permit the surviving partner to buy out the interest of the deceased or retiring partner. A partnership may have great access to equity capital from retained earnings or from contributions of capital by new or existing partners from nonpartnership sources.

d. The licensee is a person who comes on the premises with consent either expressed or implied. The possessor is liable for injuries caused by wanton or willful acts or by intentional harm or by gross negligence. The licensee assumes the risk of injury from normal activities or the normal condition of the premises when he enters the land. Examples of licensees are social guests or recreationists using the land with permission without charge. An occupier owes no duty of inspection or affirmative care to make the premises safe for a licensee; however, there is an obligation on the part of the occupier to exercise reasonable care for protection of the licensee.

e. The main problem with the present system is time. The present system is slow compared to the proposed upgraded system and other systems in schools in the area. The Data Center is getting the work done. However, if any new jobs are required, overtime will become a necessity. Also with the increasing size of data processing classes, waiting for three hours to get on the computer is not uncommon.

f. Mass recreation is heavily influenced by the distance or the time people must travel to reach the recreation site. In most areas of Ohio, weekend users will not travel more than two to three hours of driving time. For day use, a one-hour drive is generally considered to be the maximum time users will spend traveling to a recreational facility. While people like to get to the recreational area as quickly as possible, the return trip is probably more important. Long, difficult drives home usually reduce the chances for frequent return visits.

g. Eggs in refrigerated storage overnight result in an improved ability to identify shell damage by candling. About twice as much shell damage is detected after overnight storage as is found prior to storage. This may account for some of the excessive cracking found at times in retail packs—when the eggs were okay when they went over the candlelight at the time of cartoning.

h. An electric shear-type debeaking instrument usually is used to debeak day-old chicks. One of this type removes a portion of the beak with a quick stroke of the blade but does not cauterize the end of the beak. An instrument that cauterizes the cut to prevent excess bleeding is used to debeak older pullets.

i. When considering the price per acre, the person interested in purchasing rural property is most concerned about the size of the monthly payments and whether his income can support the new purchase. He considers the purchase of land in the same way he considers the purchase of other durable consumer goods, such as cars, travel trailers and boats. As a consumer, the buyer of land is valuing certain characteristics of the property on the basis of his anticipated future yield of satisfaction. While he may hope economic returns will someday accrue, he is more concerned that the land fits his mental image of a ranch or weekend hideaway.

j. The characteristics of anaplasmosis are influenced by the age of the animal, the virulence of the anaplasma parasites, and the amount of exposure. The infection can cause a rise of body temperature, followed by a progressive anemia or loss of red blood cells. Be on the lookout for an especially rapid form of anaplasmosis, in which cattle may die within a few hours after the onset of infection. This generally is called the peracute form. In addition to anemia, milk flow is suspended, very rapid respiration is noted, and affected animals may exhibit irrational behavior or signs of nervousness.

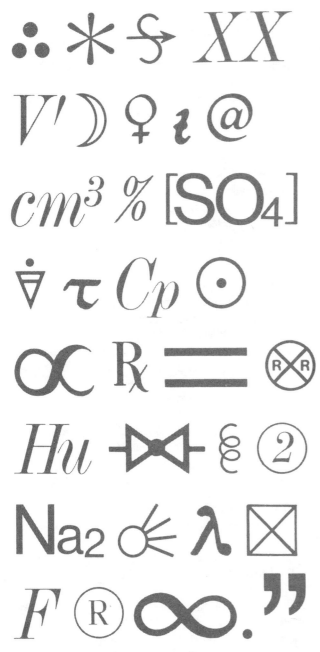

8.1 Introduction

Erika, aged ten, is looking for good books to read at the library. She reaches for a book with an attractive cover design, then she flips through the book to evaluate the pictures inside. If the content and style of the pictures please her, she will want to read the book. However, if the pictures appear dull or old-fashioned, she rejects it immediately and hunts for another. It may be an excellent book, but without the catchy cover and pictures, it remains on the shelf unread.

When Erika finds a book with an attractive cover and interesting pictures, she looks more closely; however, if the book's pages are dark with tiny print, close lines, and few breaks between paragraphs or sections, Erika puts it back. She appreciates a book with a pleasant format, including plenty of white space between the lines, extra spacing between paragraphs, and generous paragraph identations, because this shows her that the material is readable. She knows that heavy, dark, massed print and unbroken paragraphs signal material difficult for her to follow. She hunts for a book with short, snappy conversations and fast-moving paragraphs, to go along with the interesting pictures and cover. The book with an unattractive format also collects dust on the shelf.

Of course, an older reader pays closer attention to the content of a book. Fourteen-year-old Inge uses more sophisticated techniques in selecting a book in the library. She reads a few paragraphs at the beginning of the book and perhaps samples a few pages here and there before she decides which book to bring home. But still, even Inge subconsciously judges books much as Erika does, according to physical attractiveness and format.

What does all this have to do with your technical report writing? Very simply, adults also react negatively to unattractive, difficult text. Consider how you react to your own textbooks. You cringe at a heavy textbook with a sick color on the cover, pages grey from print bleeding through the thin paper, and overloaded with tight, heavy black print, and with only skimpy margins. Such textbooks very definitely have a negative psychological effect. You much prefer the interesting-looking book with an attractive cover, clear and easy-to-read print, wide margins, well-labeled pictures and diagrams, and effective headings.

The readers of your technical reports will want these same things, so make your technical reports a pleasure to look at and easy to read.

This chapter will show how to transform the mental concept of a technical report into an actual physical design on paper. You have already learned how to create maximum clarity and efficiency in your writing, to gather and select materials appropriate for your audience and purpose, to arrange these materials in an effective order, to write a well-organized rough draft, and to revise the sentences and paragraphs. Here, you learn to transform the rough draft into a prototype draft. We focus on the physical design of your writing and on the diagrams, graphs, and charts that illustrate it.

Because the draft you hand over for final typing usually will be produced without significant changes in the text, layout, or visual aids, you must make important design decisions when you prepare the prototype draft. If you hand in a sketch of an inadequately labelled visual aid, the finished product will also be inadequately labelled, no matter how skillfully it has been redrawn. If you hand in a prototype draft without adequate headings, there will be no headings added to the finished product. If you hand in a prototype without numbering, there will be no numbers in the final copy. If you hand in a prototype with inadequate white space, the final copy will have an unattractive format. It is not the role of the illustrator, editor, or secretary to make your design decisions; you must make these decisions yourself.

After you have read this chapter, you should be able to:
1. Use white space to emphasize the organization of your reports and to assist your readers.
2. Number the parts of your reports to show the organization and thereby give your readers a built-in outline.
3. Use appropriate headings to indicate the topics, enabling your readers to skim and to find sections they need.
4. Plan your visual aids effectively, thereby simplifying and clarifying ideas for your readers.

The chapter is divided into two major segments, one on layout and the other on visual aids. The segment on layout discusses the use of white space, numbering, and headings; the segment on visual aids covers the design and placement of charts, graphs, and diagrams. Your mastery of these physical design concepts will improve your ability to prepare effective reports.

8.2 Layout

Technical writing should be easy to read. Most readers will page through a report quickly, reading a bit here and there as they hunt for the parts that specifically concern them. If you have not clearly separated and labelled the different parts of the design, your readers will not be able to skim efficiently. They may have to grope along on their own or read the whole report straight through, which can lead them to conclude that the report is badly planned. Therefore, you must not only put your report in order, but give it the appearance of order as well. This is what white space, headings, and numbering accomplish. They make it possible for readers to see the conceptual design of the report at a glance, making its strengths clear.

8.2.1 White Space

A page that is dark with print is uninviting and has a poor psychological effect on the reader. You must use white space to relieve the eye and to indicate the relationships sections have to one another. Begin by considering the places where white space occurs. It is found at the margins of a page and between lines, paragraphs, and sections. The use of white space in all these places must be planned.

8.2.1.1 Margins
First, white space separates the text from the edges of the page; this white space makes up the margins. In Fig. 8-1 we have presented material from technical reports in a variety of formats to illustrate both effective and ineffective use of margins. Version A, the page without adequate margins at the sides, top, and bottom, has an oppressive, overwhelming effect. The text in Version B, on the other hand, looks skimpy. It suggests that the writer was trying to stretch the material out, with the result that the reader feels cheated. Version C uses margins effectively. The margins are generous enough to permit the report to be read easily, to permit annotation, and even to permit the report to be bound. Although generous, the margins are not excessive.

8.2.1.2 Paragraph Indentation.
Second, white space is used in the block indentation of paragraphs; this is use of horizontal white space. Paragraph indentation makes the organization clear to the reader by making the beginning of the paragraphs more readily apparent. The indentation as a whole as well as the separation of paragraphs enable the reader to see different levels of generality

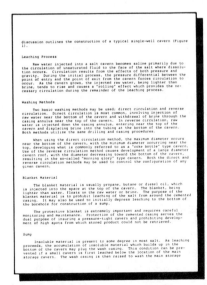

A. Page with Inadequate White Space   B. Page with Too Much White Space   C. Page with Effective Use of White Space

Fig. 8-1
Effective and Ineffective Use of White Space in
Margins

in your text. Notice how the paragraph indentation in Fig. 8-2 helps you see which are the important ideas and which are the supporting details. Notice the complete consistency of the pattern of indentation: double space between paragraphs and five spaces of indentation in the first line of each paragraph.

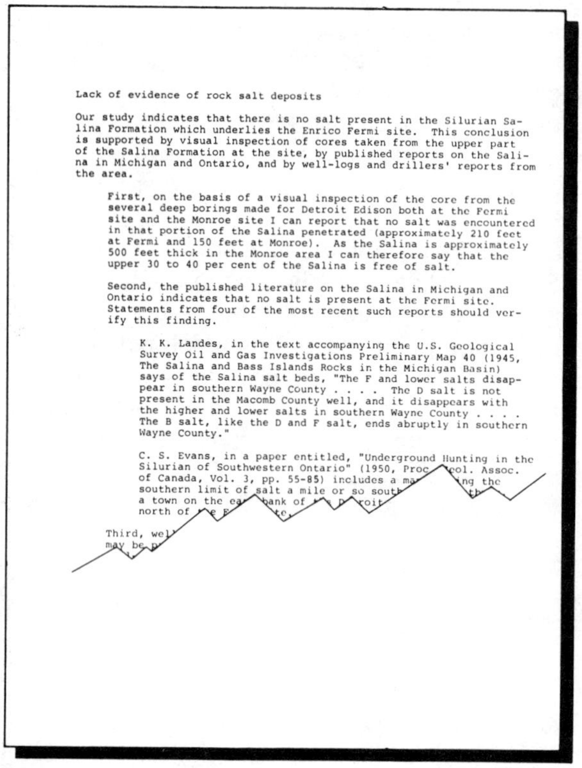

Fig. 8-2
Use of Block Paragraph Indentation to Signal Different Levels of Generality

8.2.1.3 Line Spacing

Third, there is white space between the lines and paragraphs on the page. Usually reports are typed in single-spaced format, with a double space between paragraphs. Sometimes they are uniformly double-spaced, that is, with an equal amount of white space between all lines and paragraphs. Fig. 8-3 illustrates these alternatives. Either of these patterns is acceptable, although the first is preferred. The single-spaced pattern, Version A, clearly permits the reader to see the divisions within a large block of material more easily than does the double-spaced pattern, Version B. The additional line of white space between the single-spaced paragraphs strongly clarifies the organization of your text for the reader.

8.2.1.4 Paragraph Spacing

Finally, white space is used to separate the units of a report. In Fig. 8-4, the white space tells the reader that the first paragraph on the page forms one unit of thought and that the second and third paragraphs together form a second unit. The additional white space between the two units tells the reader that the relationship between the first and second paragraphs is different from the relationship between the second and third paragraphs. Even without numbering or headings, this use of white space helps the reader.

Do not underestimate the effectiveness of white space. A speaker can often quiet a noisy audience just by standing silently in front of them. White space on a page of print is like silence, in that it can communicate very effectively. On a page full of print, a block of unprinted lines stands out immediately. The reader knows that the space signals some change. It means that the division, segment, unit, or paragraph is ending and that another one is beginning. It indicates that the report is moving from one level of importance to another or from one unit of thought to the next. Therefore you must plan your white space as carefully as your words.

8.2.2 Numbering

Like white space, numbering signals the

discussion outlines the construction of a typical single-well cavern (Figure 1).

Leaching Process

Raw water injected into a salt cavern becomes saline primarily due to the circulation of unsaturated fluid to the face of the salt where dissolution occurs. Circulation results from the effects of both pressure and gravity. During the initial process, the pressure differential between the point of entry and the point of exit from the cavern forces circulation to occur. As the cavern grows, the injected raw water, being lighter than brine, tends to rise and causes a "rolling" effect which provides the necessary circulation during the remainder of the leaching process.

Washing Methods

Two basic washing methods may be used; direct circulation and reverse circulation. Direct circulation is most common, involving injection of raw water near the bottom of the cavern and withdrawal of brine through the casing annulus near the top of the cavern. In reverse circulation, raw water is injected down the casing annulus, entering near the top of the cavern and displacing brine into the tubing at the bottom of the cavern. Both methods utilize the same drilling and casing procedures.

When using the direct circulation method, the maximum diameter occurs near the bottom of the cavern, with the minimum diameter occuring near the top, developing what is commonly referred to as a "coke bottle" type cavern. Use of the reverse circulation method causes development of a large diameter cavern roof, with the diameter decreasing toward the bottom of the cavern, resulting in the so-called "morning glory" type cavern. Both the direct and reverse circulation methods may be used to control the configuration of any given cavern.

Blanket Material

The blanket material is usually propane, butane or diesel oil, which is injected into the space at the top of the cavern. The blanket, being lighter than brine, floats on the raw water or brine. The purpose of the blanket material is to prohibit leaching of the salt from around the cemented casing. It may also be used to initially depress leaching to the bottom of the borehole for construction of a sump.

A. Single-Spaced Format Is Visually Effective

discussion outlines the construction of a typical single-well cavern (Figure 1).

Leaching Process

Raw water injected into a salt cavern becomes saline primarily due to the circulation of unsaturated fluid to the face of the salt where dissolution occurs. Circulation results from the effects of both pressure and gravity. During the initial process, the pressure differential between the point of entry and the point of exit from the cavern forces circulation to occur. As the cavern grows, the injected raw water, being lighter than brine, tends to rise and causes a "rolling" effect which provides the necessary circulation during the remainder of the leaching process.

Washing Methods

Two basic washing methods may be used; direct circulation and reverse circulation. Direct circulation is most common, involving injection of raw water near the bottom of the cavern and withdrawal of brine through the casing annulus near the top of the cavern. In reverse circulation, raw water is injected down the casing annulus, entering near the top of the cavern and displacing brine into the tubing at the bottom of the cavern. Both methods utilize the same drilling and casing procedures.

When using the direct circulation method, the maximum diameter occurs near the bottom of the cavern, with the mini-

B. Double-Spaced Report Obscures Organization of Report

Fig. 8-3
Effective and Ineffective Use of White Space Between Lines and Paragraphs

different parts of a report and makes the report structure clear. For that reason, you should routinely use it in most technical writing.

The numbering in a report effectively indicates where the different parts begin, defines the relationships among the parts, and clarifies the levels of generalization. In Version A in Fig. 8-5, the numbering illustrates the structure and complements the white space. The numbering in Version B indicates the levels of generality and relationships among paragraphs, even without indentations.

The numbering system also serves as a guide. Because all the subdivisions of a concept are connected by the numbering system, the reader can find any particular section easily. Notice, by the way, that this text uses only

discussion outlines the construction of a typical single-well cavern (Figure 1).

Leaching Process

Raw water injected into a salt cavern becomes saline primarily due to the circulation of unsaturated fluid to the face of the salt where dissolution occurs. Circulation results from the effects of both pressure and gravity. During the initial process, the pressure differential between the point of entry and the point of exit from the cavern forces circulation to occur. As the cavern grows, the injected raw water, being lighter than brine, tends to rise and causes a "rolling" effect which provides the necessary circulation during the remainder of the leaching process.

Washing Methods

Two basic washing methods may be used; direct circulation and reverse circulation. Direct circulation is most common, involving injection of raw water near the bottom of the cavern and withdrawal of brine through the casing annulus near the top of the cavern. In reverse circulation, raw water is injected down the casing annulus, entering near the top of the cavern and displacing brine into the tubing at the bottom of the cavern. Both methods utilize the same drilling and casing procedures.

When using the direct circulation method, the maximum diameter occurs near the bottom of the cavern, with the minimum diameter occuring near the top, developing what is commonly referred to as a "coke bottle" type cavern. Use of the reverse circulation method causes development of a large diameter cavern roof, with the diameter decreasing toward the bottom of the cavern, resulting in the so-called "morning glory" type cavern. Both the direct and reverse circulation methods may be used to control the configuration

Blanket Material

The blanket oil, which is in The nk b

Fig. 8-4
Use of White Space to Signal Units of Thought

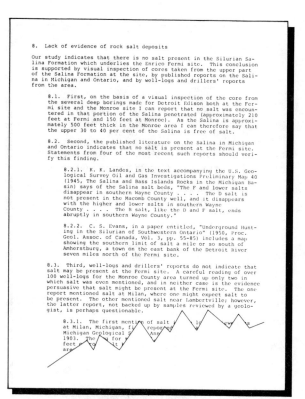

A. Numbering Complements Use of White Space

B. Numbering Alone Indicates Levels of Generality

Fig. 8-5
Use of Numbering to Signal Levels of Generality and Relationships

Arabic numbers (8., 8.1, 8.1.1), rather than the traditional system of Roman numerals, letters, and Arabic numerals (VIII, A, 1). The all-Arabic numbering system is becoming increasingly common; a form of this system is required in most Federal government reports.

8.2.3 Headings

Headings are the best way to help the reader who is in a hurry. They are a table of contents built right into the text, helping your readers quickly find what they need from your report. Headings also give your readers an outline of what is in the parts of the report they do not even read. However, headings must be well designed, that is, they must be frequent and as informative as possible.

Headings tell the reader the subject of each unit. The more subdivisions and headings there are, the clearer the structure is. Do not be afraid to carry headings and subheadings to second and third levels of subdivision. The headings in Fig. 8-6 make the structure of the unit clear at a glance. They signal that there are two levels of subdivision within the level identified by the primary heading. The different levels of headings are distinguished by formatting and white space.

Weak, vague, or general descriptive headings do not help the reader. Terms such as "Introduction," "Discussion," "Body," and "Conclusions" do not convey any precise meaning to the reader: they provide no clue to the real content of a specific section of the report. Such headings are the refuge of the lazy writer. Always use meaningful headings and subheadings, such as:

Setup of the test equipment
Experimental procedure
Comparison of predicted and measured results
Cost estimates
Effects of induced voltages

These headings are particular: they provide definite information about the content of the section. An indication that the headings give the readers a clear idea of the actual content of a particular report is that they cannot be transferred unchanged to another report; they are not interchangeable, all-purpose headings. This specific information enables your readers to read efficiently. Use informative headings as much as possible, and use as many as you need to make your topic clear.

These are the elements to consider in planning your layout: white space, numbering, and headings. Although these devices cannot impose order on a disorganized report, they can help to make the report's design more readily apparent to your readers. Remember, it is important that the report not only is designed well, but also that the report appears to be designed well.

8.3 Visual Aids

Visual aids in a technical report are the equivalent of pictures in a book. They help the inexperienced reader understand the concepts; they give the experienced reader a great deal of information efficiently. A single diagram or graph is an effective way to make information clear to your readers and can save you many words of text. A half-page wiring diagram of a stereo tape deck, for example, tells the reader how the tape deck provides input to the amplifier and speakers, and it does so much more efficiently than if the information were all in words. How many words would it take to provide all that information? How would you organize it to make it completely clear and accessible to your reader? In this case, as so often, the diagram does a better job than is possible with words alone.

In order to be effective, visual aids must be liberally used, well placed, and well designed. A well-designed visual aid is as simple as possible, is clearly labelled and captioned, is explained in the text, and uses any appropriate scales.

8.3.1 Liberal Use of Visual Aids

A simple diagram, graph, or illustration can clarify your ideas and convey information to your readers better than words. However, for the visual aid to work, the writer must decide what information lends itself to visual presentation and must take the time to design the visual aid carefully. Fig. 8-7 illustrates how a visual aid replaces words. Although the writer certainly had to put out some effort designing the aid, the economy of space in the report make this effort worthwhile. The visual explanation of creating a cavern in natural salt deposits for inexpensive storage of hydrocarbon fuels is effective. Words by themselves probably could not make the process clear in the same amount of space. The picture and words together are effective.

When you write reports, keep the readers in mind. Sketches, cutaway drawings, photographs, flow charts, bar graphs, and tables all cost you some effort, but they make your readers' task easier. For that reason, you should use visual aids frequently.

Plymouth Rehabilitation Center
OCCUPATIONAL THERAPY

INITIAL EVALUATION

Name: Mildred De Forouw   Unit: Senior West   Date: July 19, 1979

Examiner: Lori Dostal

SUMMARY OF EVALUATIONS

The patient's initial Occupational Therapy evaluation was completed in December, 1978. The patient was found to be alert, talkative and friendly. Her physical and mental states are fairly good, but some assistance will be needed in some activities.

An O.T. program including ADL evaluation has been prescribed, as well as training in precautions for maintaining safety in wheelchair transfers. In addition to these treatments, an O.T. workshop and/or floor activities will be beneficial to the patient. The entire staff should help by encouraging her to participate in social activities.

ANALYSIS

MOTIVATION

The patient's motivation toward socialization is good. She likes to talk to people and be with others. Her motivation toward work and recreational activities is fair.

ATTITUDE

The patient has a good attitude toward herself, family, peers, and staff. She also has a good outlook on her disabilities and her environment.

Fig. 8-6
Effective Use of Headings and Subheadings to
Make Structure Clear

MENTAL FUNCTIONING

The patient's mental functioning is good, and she follows directions
well. She is alert, talkative, and cooperative. Her communication
skills are fine. She enjoys watching TV and prefers reading books
with large print.

PHYSICAL FUNCTIONING

Some assistance is needed with ambulation, as the patient's legs
are not very strong. Her upper extremities' range and strength
are within normal limits, although her neck rotation is limited.
Only minimal assistance is needed with dressing. The patient's
coordination and grasp are good, and her sensory abilities are
normal.

ASSESSMENT OF POTENTIAL

The overall assessment of her potential is fair to good, because
she is cooperative and friendly. She is motivated and functions
well, but she needs persuasion to become involved in work and
activities.

GOALS

SHORT-TERM

Our goals for the patient, on a short-term basis, are for her to
establish independent activities of daily living. We also expect

Fig. 8-6
Effective Use of Headings and Subheadings to
Make Structure Clear

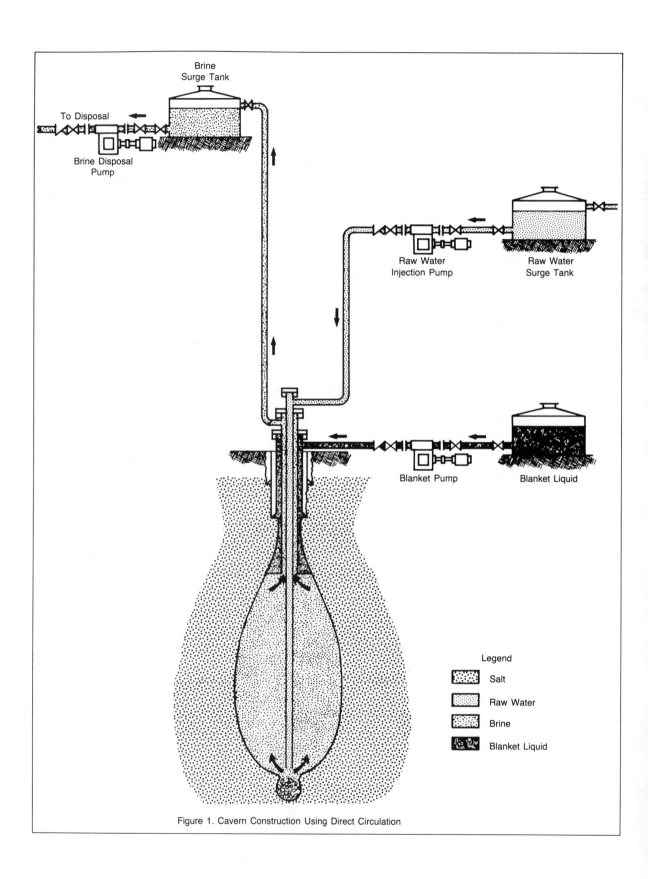

Figure 1. Cavern Construction Using Direct Circulation

Fig. 8-7
Visual Aid More Effective Than Words

The Technician as Writer

As you increase your skill in report writing, you will make the various types of illustrations an integral part of your report and will mentally begin to design them as you prepare the first draft. The beginner, however, should review his or her report during the editing procedure and insert an illustration wherever it can supplement, clarify, or replace part of the text. Think of your readers, not of yourself. You will find that the time spent preparing the visual aids is well worth it to you in the long run.

8.3.2 Effective Placement of Visual Aids

Unfortunately, many report writers place all their illustrations, graphs, and tables in the appendix. This made some sense at one time, because reproducing visual aids used to be difficult and expensive. It used to be easiest to make all visuals full-page size and to group them all together at the back of the report. Today, however, any copier can reproduce your visuals, and many can even reduce them so that they fit right into the text where the readers need them. Fig. 8-8 presents a visual aid that was attached to the end of a report rather than broken up and placed where the drawings would have been of more use to the reader.

To see the effect of putting the visual aids in the appendix, consider how you read. Suppose you pick up a report and begin to read. You come to a passage: "See the diagram in the appendix." You flip to the appendix, because the prose does not make sense without the diagram. Then you flip back to the text, trying to put the diagram and the text together in your mind. Then you come to a page that has four or five references to appendix visual aids. You have only five fingers with which to keep the diagrams handy while you try to interpret the text. After a few attempts to keep your fingers and the visual aids straight, you are likely to get irritated and quit looking. The visual aids, no matter how useful and well designed, are no longer doing you any good.

If you have a lot of visual aids, you must decide which to put in the text and which in the appendix. If you put them all in the text, you'll separate sections of the report text so far that the reader will miss some connections. If you put them all in the appendix, the reader may not refer to them. To help your readers, put the most important visual aids, the ones you are sure they will really need, in the text. Place them on the page where you discuss them or, if that is not practical, on the page immediately following.

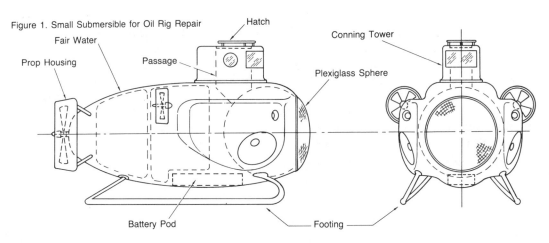

Figure 1. Small Submersible for Oil Rig Repair

Hatch — Fair Water — Prop Housing — Passage — Conning Tower — Plexiglass Sphere — Battery Pod — Footing

Fig. 8-8
Visual Aid Attached to a Report Rather Than
Reduced or Subdivided to Illustrate the Text Directly

8.3.3 Good Design of Visual Aids

In planning the report, your prime consideration is your audiences' needs. This is also true for planning your visual aids. You must suit your visual presentation to the readers' abilities, knowledge, interests, and needs. Do they have the background to understand a certain diagram? If not, how can you simplify that diagram so that your audience can understand it?

A word of caution is in order here. Be realistic about your audience. A complicated schematic drawing may be easy for you and other specialists to understand and might be appropriate for some reports. The schematic drawing shown in Fig. 8-8 probably makes sense only to specialists in oceanographic technology. But what about the other audiences for this particular report? Will the drawing help them understand the material or will it serve only to confuse them further? Do the front and side view drawings clarify or obscure matters for the readers? For people without a technical background and even for most technical readers, a simplified isometric view, as in Fig. 8-9, would be much more effective. The isometric view sacrifices technical detail, but the nonspecialists can understand it.

An architect's sketch of how a finished house will look, while not accurate enough for a carpenter's purposes, nonetheless conveys very useful information. It certainly does not replace the blueprint, but it is more meaningful to the nonspecialist. Because many reports have both specialist and nonspecialist readers, you may need to provide both types of visual aids. Put the technically complex visual aids in the appendix; keep the simplified ones with the text. The artist's three-dimensional sketch (Fig. 8-10) will tell most people much more than a

technician's drawing, and you can probably do such a sketch for your own reports.

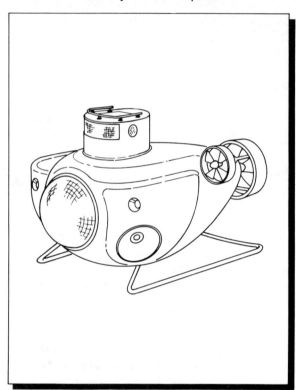

Fig 8-9
Visual Aid Simplified for the Nonspecialist: Isometric Projection

How do you know whether a visual aid is suitable for your report's audience? Ask a friend who is a nonspecialist to look at it to see if he or she understands it. Generally, as you become more specialized in a field, you tend to overestimate the understanding of the nonspecialist. Always choose your visual aids with your audiences' needs firmly in mind.

Remove all unnecessary clutter from your visual aids, in order to help your readers see the really important points. Even removing a few lines from a graph or unneeded columns from a table may enable the reader to read it more easily. In the table in Fig. 8-11, Version A, the numbers obscure the relevant information; in Fig. 8-11, Version B, the table communicates the essential information.

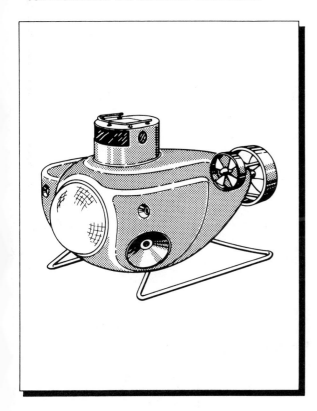

Fig. 8-10
Visual Aid Simplified for the Nonspecialist: Artist's Sketch

The reader may misinterpret unlabelled visual aids. For instance, what does Fig. 8-12 represent? Is it a map of Transylvania or the cellular structure of an African violet? Actually, it illustrates the cracking or "alligator" effect that occurs when several heavy coats of paint are applied without sufficient drying time between them, but no one would have been able to tell this from the figure as it stands.

Each visual aid must be clearly captioned and labelled, so that readers know what they are looking at and what they are expected to learn. You cannot assume that your readers will all see what you want them to, unless you tell them what it is.

Your readers need complete information about every visual aid in order to be able to interpret it. The diagram or picture should have a caption stating its subject and purpose; all the parts should be clearly labelled; and there should be a figure number identifying it.

Many of your readers will leaf hurriedly through your report, looking for instant information. Visual aids and tables that are carefully captioned and precisely labelled give those speeding readers accurate information. Don't expect the reader to stop and read through the text for clues to a visual aid's purpose. Notice how directly the figure from the specimen report in Fig. 5-6 communicates when the caption and labels are added (Fig. 8-13). The purpose of visual aids is to clarify and provide information, not to force reluctant readers to read the text.

The reader must be told what a diagram or graph represents, what is to be seen in it, and why it is there. Unless you put into words what you want the reader to see, you cannot be sure the visual aid will be understood as you intend. In Fig. 8-14, the words and the simple illustration together achieve brevity and clarity. In Fig. 8-15, the words make the point, and the bar graph clarifies the meaning of the statistics given in the text. The same is true in Fig. 8-16.

The text must direct the reader to look at the visual aid from some particular point of view. Bar, pie, and other kinds of graphs help the reader understand the implications of statistics,

```
HAND RESPONSE EXPERIMENT RESULTS

    The results of the experiment in which we monitored the right
hand skin temperature during thermal stimulation of the left hand
are listed in Table 3.  Two tests were performed for a +15°C step
on the left hand; two tests were performed for a -15°C step on the
left hand; and one test was run for a -30°C step.  The averages of
the water and skin temperatures both before ("B") and after ("A")
the thermal step the the left hand are also listed.

Left Hand Step Input of:
    +15°C                  -15°C                  -30°C

Right Hand:
Bath   Skin   Skin    Bath   Skin   Skin    Bath   Skin   Skin
Temp   Temp   Temp    Temp   Temp   Temp    Temp   Temp   Temp
         B      A               B      A               B      A

Test 1                 Test 3                 Test 5
21.3   24.6            22.1   23.1            22.4   23.2
21.2   24.6            22.1   23.1            22.3   22.8
21.4   24.5            22.2   23.1            22.4   23.1
21.4   24.3            22.1   23.2            22.3   22.9
21.4   24.3            Ave-   Ave-            22.4   22.9
21.7   24.3            22.13  23.13           22.4   22.9
21.8   24.1                                   Ave-   Ave-
21.4   24.3            22.2          23.4     22.37  22.97
Ave-   Ave-            22.1          23.1
21.46  21.38          22.2          23.1                    22.9
                       22.2          23.1                    22.9
21.7          23.9    22.2          23.2                    22.9
21.6          24.1    Ave-          Ave-                    22.8
21.6          23.9    22.18         23.18                   22.9
21.7          23.9                                   Ave-   22.9
21.7          23.8    Test 4                         22.38
21.7          23.8    22.1   22.9
Ave-          Ave-    22.2   23.0
21.67         23.90   22.2   23.0
                       22.2   22.9
Test 2                 22.2   22.9
21.7   23.7           Ave-   Ave-
21.5   23.7           22.18  22.94
21.7   23.6
Ave-   Ave-           22.2          22.9
21.36  23.67          22.2          22.8
                       22.3          22.9
21.7          23.7    22.5          22.8
21.8          23.5    22.3          23.0
21.7          23.6    22.2          23.0
Ave-          Ave-    22.2          23.0
21.73         21.60   Ave-          Ave-
                       22.29         22.94
```

A

```
HAND RESPONSE EXPERIMENT RESULTS

    The results of the experiment in which we monitored the right
hand skin temperature during thermal stimulation of the left hand
are listed in Table 3.  Two tests were performed for a +15°C step
on the left hand; two tests were performed for a -15°C step on the
left hand; and one test was run for a -30°C step.  The averages of
the water and skin temperatures both before ("B") and after ("A")
the thermal step the the left hand are also listed.
```

	Left Hand Thermal Stimulation								
	+15°			-15°			-30°		
Test	Bath Temp	Skin Temp Before	Skin Temp After	Bath Temp	Skin Temp Before	Skin Temp After	Bath Temp	Skin Temp Before	Skin Temp After
1	21.46	24.38							
	21.67		23.90						
2	21.36	23.67							
	21.73		21.60						
3				22.13	23.13				
				22.18		23.18			
4				22.18	22.94				
				22.29		22.94			
5							22.37	22.97	
							22.38		22.90

```
    TABLE 3:  Experimental data for the right hand skin temperature
              response experiment

DISCUSSION OF RESULTS OF SKIN-TEMPERATURE RELATIONSHIP

    Our results do not agree completely with any of the previous
investigations, as shown in Fig. 1.  In particular, our results
indicate that the temperature difference between the skin and the
ambient environment is almost constant as the hand is cooled, and
that the hand exhibits a hysteresis effect in its response to...
```

B

Fig. 8-11
Ineffective and Effective Versions of a Table in a Report

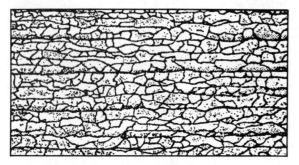

Fig. 8-12
Visual Aid Without Caption to Identify Subject

but these implications must also be discussed in the text. The visual aid and the text work hand in hand to express ideas efficiently and effectively. This built-in redundancy is necessary to address both people who are reading superficially and those who are reading carefully. Obviously, the visual aid should be located as close to its discussion in the text as possible.

If a visual aid misleads the reader, it is worse than useless. In Fig. 8-17, the two graphs have different scales, and one has a suppressed zero (a zero below the bottom of the scale). Therefore, they imply to the casual reader that the runoff of the Green River is greater than the annual precipitation. Fig. 8-18, with the two graphs redrawn on the same scale, one superimposed on the other, conveys the correct impression to the reader. As you can see, the runoff is only a tiny fraction of the precipitation. That was certainly not the impression created by the original graphs.

Effective layout of your text and effective use of visual aids can often make the difference

The Technician as Writer

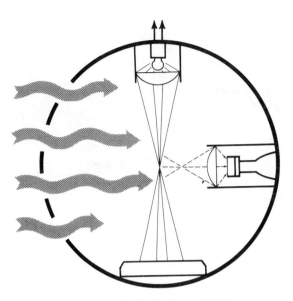

A. Schematic Without Captions and Labels

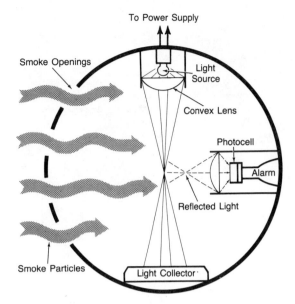

B. Schematic with Captions and Labels

Fig. 8-13
Use of Captions and Labels for Direct
Communication

between a mediocre report and an excellent
one. A report consists of more than words; it is
a physical object. You must consider layout
and visual aids just as carefully as basic
structure and paragraphing. If your original
concept is not well implemented, the idea of
your report will be clear only to you. If original
design is clearly executed and supplemented
by effective use of white space, numbering,
and headings and by well-designed visual aids,
your readers will understand the report as you
intend them to.

Bait Boxes

Bait should be placed where only rats and mice can feed on it.
The following drawing shows a simple method of protecting bait.

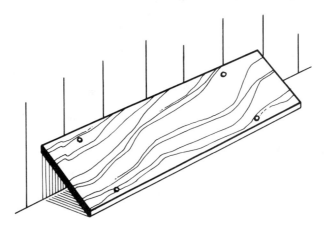

Use about ½ pound of bait at each station. Place baits where
rats and mice are known to feed along walls inside and outside of
buildings, in dark corners, under floors, in attics and under stair-
ways. The entire premises should be baited at the same time. The
bait may be placed in a shallow dish or in a small paper sack. Paper
sacks should be slit open at the side so that some of the bait will
spill out. This hastens feeding. Sacks should be nailed to the
floor or wall to prevent rats from dragging them away.

There is no hard and fast rule for the number of baits to put
out. For small buildings with a few rats, 2 pounds should be
enough. If there are many rats, 3 or 4 pounds may be needed. In
large buildings with a few rats, 2 pounds may be enough; but if
there are many rats, 5 pounds may be required before completing the
job. The average for a Texas farm is about 5 pounds.

At first put out many small baits to see where the rats and
mice prefer to eat and then move uneaten baits to these places. If
not enough bait was made up at first, a new batch should be prepared
before containers are entirely empty. Where there is a source of
reinfestation from other areas, such as dump grounds or nearby
infested farms, baits should be kept out at all times for new rats
as they come in. Some anticoagulant baits may become weevilly or
rancid. These should be replaced with fresh bait.

In addition to the dry form, a soluble material is available
for use in a water bait. One packet is sufficient for 1 quart of
water which is exposed in baby-chick water fountains or similar
containers. In dry surroundings this is particularly effective.

Questionnaire responses indicated that laboratory test reports comprised the bulk of reports produced; in-house proposals, trip reports, progress reports, and responses to government inquiries made up the rest.

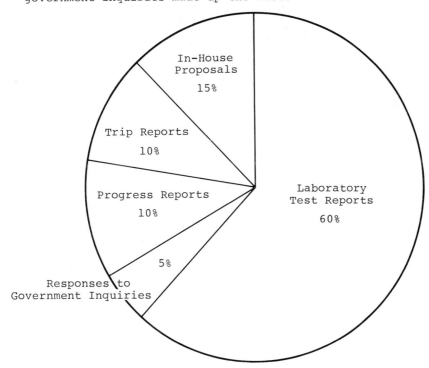

Figure 2-1. Percentage of Reports By Types At Automobile Proving Grounds, Lab Performance Test Section.

Fig. 8-15
Circle Graph Introduced by Text

187

Although the trend in the balance between the sexes is in the same direction in Texas and in the nation as a whole, there were more women than men as early as 1950 on a nationwide basis. The sex ratio for the nation in 1970 continued to be slightly lower than in Texas, the ratios showing 94.8 and 95.9 males for every 100 females, respectively.

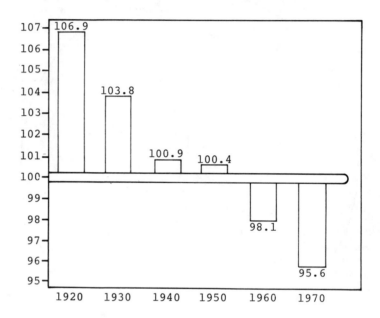

Figure 13. Number of males per 100 females in the Texas population, 1920-1970.

Fig. 8-16
Bar Graph Introduced by Text

The Technician as Writer

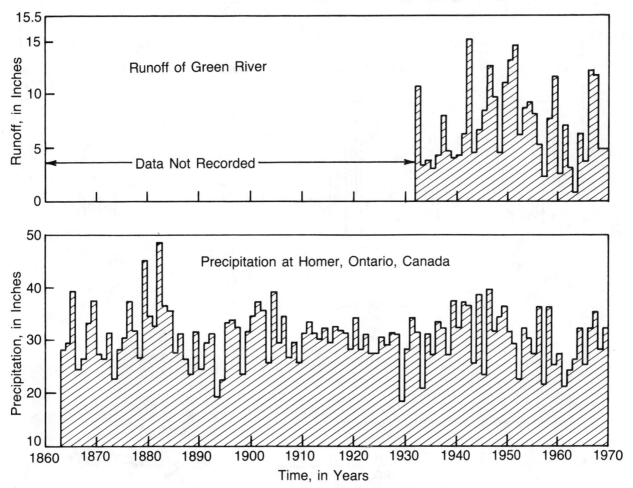

Figure 7: Annual Precipitation and Runoff

Fig. 8-17
Visual Aid Which Distorts Because of Different
Scales and Suppressed Zero

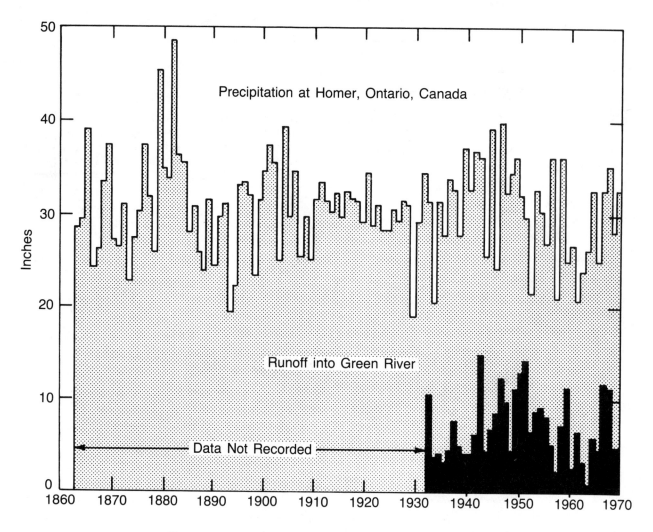

Figure 7: Annual Precipitation and Runoff Relationship

Fig. 8-18
Visual Aid with Explicit Comparisons Made on the
Same Scale

The Technician as Writer

Exercises

Apply the principles explained in this chapter to the following examples. The visuals will be easy to prepare because the data, although based on real facts, has been rounded off or approximated and simplified.

A. For the following material from a technical article,[1] convert each of the tables or other items providing statistical information into an appropriate visual aid, formatting each one effectively to illustrate the concepts about work. Your instructor may select from these, rather than having you do the entire set. Afterwards, compare your visuals with the ones your instructor will provide. Of course, your responses will differ, but those differences will provide a basis for productive class discussion of how visual aids are best designed.

Opposing Views About Work
There are two extreme opposing views of work—the "efficiency" view and the "humanistic" view. They are well characterized in the classic fable about the grasshopper and the ant. The efficiency view emphasizes the necessity of work in order to obtain the prerequisites for survival: food, clothing, and shelter. Work is necessary, but it is also onerous, at least to some degree; therefore, work time should be minimized and leisure time should be maximized. Work time can be minimized through the specialization of skills, which allows higher output rates to be achieved per unit of work input. With technically planned production systems, in other words, higher production efficiency can be attained. The result is increased social benefits in the form of more leisure time, lower costs for goods, services available in increasing quantities, and a heightened standard of living, thanks to the efficient utilization of human and economic resources. In recent years, this view of work has gained more and more critics; they emphasize the prevailing tendency to view production efficiency as a technological problem, with workers

considered as cogs in a larger mechanistic environment. In such an environment, they point out, work is reduced to simple repetitive tasks that require little creativity and provide little satisfaction or sense of achievement.

According to the opposing "humanist" view, work is necessary but need not be onerous. In fact, the proponents of this view argue that since most of us spend a substantial part of our lives working, work should be not only productive but satisfying, safe, and socially and intellectually stimulating. Because the nature of our work colors our perspective of the world, indeed influences our longevity, unpleasant work should be eliminated. Since profit-oriented technological industries are impersonal entities with little concern for humanistic values, unions and governmental legislation are necessary to assure the worker's right to employment that is satisfying and safe. For individuals who are unable to obtain satisfactory sustenance, social programs are needed to guarantee health care, unemployment compensation, retirement benefits, affirmative action, and so on. The critics of this view argue that social regulations controlling the work place are incompatible with a capitalistic economic system; that they increase the cost of output and cause a loss of productivity in all countries where the regulations apply. The result is a weakened ability to compete in world markets and a subsequent loss of foreign exports and domestic jobs.

Data Related to Work[2]
The efficiency and the humanistic views of work are both extreme and naive. In order to sharpen and clarify the arguments, let us review a few of the data pertaining to those positions. All of the graphs and charts presented here are based on hard data, but many of the slopes have been smoothed in order to emphasize or clarify the basic trend.

This shows the growth of the US population and the expanding labor force from the early 1800s to date.

---

[1] These passages have been adapted from Richard C. Wilson, "Work and the Engineer," *Technium,* College of Engineering, University of Michigan, Ann Arbor: Fall 1976, pp. 14–27.

[2] In this material, we have added the outline numbers indicating exercises and have translated the visual aids of the original into tables. The words are the original author's.

1.

Year	Population in Millions	Labor Force In Millions
1800	20	—
1875	50	20
1900	80	45
1965	200	85
2000	290 (Projected)	—

In the last several decades, the characteristics of this labor force reveal a number of marked changes.

One of these changes is in the level of educational attainment of the members of the labor force aged 18 and over. By 1980, about 30% of the labor force will have completed at least one year of college, and at least 15% will hold college degrees. At least 75% will have completed four years of high school, compared to about 40% in 1970.

2A.

Year	Level of Education		
	Elementary 1-8 years	High School 1-4 Years	College 1 or More Yrs.
1960	30%	50%	20%
1970	25%	50%	25%
1980 (Projected)	20%	50%	30%

Another change is that the proportion of the labor force is greatly increasing in the 25-34 age range.

2B.

Year	Percent of Total Labor Force		
	55 or Over	25-34	35-54
1950	17%	25%	44%
1960	18%	22%	45%
1970	18%	20%	43%
1980	17%	26%	40%
1990	12%	28%	41%

A third change is that the number of female workers in the labor force is growing at a rate far exceeding that of male workers.

2C.

Year	Increase (Index: 1950-100)	
	Male	Female
1950	100	100
1960	110	120
1970	125	150
1980	144	188
1990	155	220

The labor force is experiencing an employment shift from goods-producing industries to service industries.

The magnitude of this shift is illustrated by the proportion of the US Gross National Product in each of these two industries.

3A.

Year	% of GNP Services	% of GNP Goods	GNP in U.S. Dollars (Billions)
1950	30.6%	69.4%	$294.8
1960	37.2%	62.8%	$503.8
1970	42.1%	57.9%	$974.7

This shows the accompanying shift from agriculture to manufacturing and recently to services employment.

3B.

Year	Agriculture	Mining Manufacturing Construction	Utility	Commerce	Services
1870-79	60%	20%	—	—	—
1950	20%	28%	12%	—	—
1960	14%	30%	12%	20%	20%
1969	8%	32%	12%	25%	25%

Projected to 1980, the shift is even more striking.

3C.

Year	Labor Force (Millions)	
	Services	Goods
1945	26	30
1954	34	30
1963	40	30
1972	50	33
1980	62	35

Given these major shifts in the US work scene, it is interesting to note that surveys of workers disclose no substantial shift in reported job satisfaction. Perhaps this results from the employment trend away from jobs with low educational requirements (such as those for operators, laborers, and farmers).

3D.

Year	% of Satisfied Workers	
1958	74% 21-34 years of age	UM Survey Research Center
	79% 35-44 years of age	
	77% High School	
	81% Beyond High School	

1964	87% 21-30 years of age	National Opinion Research
	93% 31-40 years of age	Center
	90% High School	
	94% Beyond High School	
1968	76% 21-29 years of age	UM Survey Research Center
	88% 30-39 years of age	
	86% High School	
	85% Beyond High School	
1973	84% 21-29 years of age	UM Survey Research Center
	92% 30-39 years of age	
	89% High School	
	91% Beyond High School	

On the other side of the coin, a number of domestic and international issues draw attention to the need for efficient work.

This shows the attempts of three economists to summarize the factors that affect productivity growth. Ignoring niceties of definitions, concepts, and assumptions, we see that all three include similar factors in their summaries. ("Technology" is not a measured effect, but rather the residual part of productivity growth not accounted for by the other factors measured.) In each case, we see that technology and capital per worker are the major sources of productivity growth, and that "labor quality" is at best a minor component.

4.

Kendrick
More Capital 18.8%
Labor Quality 9.3%
Technology 71.9%

Thurow
More Capital 29.8%
Labor Quality 30.2%
Technology 40%

Denison
More Capital 25.4%
Better Resource Allocation 9.5%
Economics of State 12.7%
Labor Quality 14.3%
Technology 38.1%

Currently there is a widespread view that the USA is losing its economic affluence and productivity leadership in part because its technology is not competitive with that of other nations. This is inferred from several factors.

The first is the large share of the US market enjoyed by engineering-manufactured imports.

5A.

Ratio of Imports to Consumption: 1970

Textiles (including apparel)	12%*	*By Value
Steel	15%**	**By Volume
Flatware	22%*	
Footwear (non-rubber)	30%*	
Leather Gloves	30%*	
Sewing Machines	40%*	
Black & White Televisions	52%**	
Amateur Motion Picture Cameras	66%*	
Radios	70%*	
Calculating Machines	75%*	
Hairworks, Toupees & Wigs	85%*	
Magnetic Tape Recorders	86%**	
35mm Still Cameras	100%**	

The second factor is the decline in the US share of world production in such major categories as automobiles and steel.

5B.

Year	Motor Vehicles (Annual Average)				
	US	Japan	West Germany	France	Other
1949-50	76%	1%	3%	10%	10%
1959-60	48%	1%	13%	10%	28%
1969-70	31%	18%	13%	10%	28%

Year	Steel				
	US	USSR	Japan	West Germany	Other
1949-50	48%	16%	1%	7%	28%
1959-60	28%	20%	9%	11%	32%
1969-70	20%	20%	18%	10%	32%

The third factor is the lagging rate of growth in output per man-hour in US manufacturing, compared with that of other nations.

5C.

Year	Index (1960 = 100)					
	US	United Kingdom	Canada	European Economic Community	Sweden	Japan
1960	100	100	100	100	100	100
1962	110	110	103	112	113	140
1964	112	113	114	120	127	140
1966	125	125	126	137	150	170
1968	126	135	138	160	175	210
1970	130	140	150	175	200	275
1972	140	157	170	190	215	315

A number of possible reasons for these effects have been suggested. Here are three of them.

One possible reason is that the US economy is

bearing a larger burden of non-goods-producing military expenditure.

5D.

	Share of GNP for Military Expenditure
Japan	0.8%
Italy	2.6%
West Germany	2.9%
France	3.8%
United Kingdom	5.6%
US	7.8%
USSR (estimated)	8.0%

Another possible reason is that the US investment in new plant and equipment (and hence the introduction of new technology into the manufacturing sector) lags behind that of its major international trade competition.

5E.

	Share of GNP for Investment in New Plant & Equipment (1968-70 Average)
US	10%
Canada	13%
France	18%
West Germany	19%
Japan	20%

A third possible reason is that the profitability of US nonfinancial corporations is decreasing as a share of GNP.

5F.

Year	US Dollars Before Taxes (Billions)			
	Total	Manufacturing	Transportation Communication and Public Utilities	Others
1958	36 (4.4% of GNP)	20	7	9
1962	44 (4.7% of GNP)	28	6	10
1966	74 (5.8% of GNP)	45	11	18
1970	63 (3.3% of GNP)	35	5	23

It is especially significant that the highest percentage increases in GNP expenditure occur in private plant and equipment, transportation (especially highways), education, environmental quality, social welfare, and health, whereas the smallest changes are in urban development, national defense, research and development, agriculture, and international aid. It is significant also that such countries as Japan have been

beneficiaries of US technology during the period 1950-70, through licensing agreements and the growth of multinational US corporations.

Now, compare your visuals and formatting with the originals provided by your instructor. Discuss with the class which visual aids are the most effectively designed. Have any of the class members surpassed the efforts of the original designer?

B. Prepare a mock-up of an effective layout of the unformatted report distributed by your instructor. Do not retype the report; just use headings and section blocks on blank pieces of paper to indicate how the report should look.

C. Suggest improvements in the specimen visual aids distributed by your instructor.

9.1 Introduction

Have you ever watched a fine cook prepare a meal? The kitchen and utensils are readied. Even before this, the dessert may have been made and put in the refrigerator along with the rinsed lettuce and vegetables. Then the oven dish is prepared and put in to bake. The potatoes are put on to boil, and the sauces are simmered. Next, the vegetables are set to steam, the rolls put in the oven to heat, and the lettuce brought out and the salad prepared. While you set the table, the vegetables are taken off the stove and garnished. Then the rolls come out, and the salad is tossed. Finally the casserole comes out of the oven. Everyone sits down at the table and is served a simple three-course dinner.

The dinner is a success. Some will compliment the sauces, others the casserole, and those with a sweet tooth will be sold on the dessert. But of course the real reason for the success will be the careful planning and preparation by the cook.

Preparing your report is like preparing a three-course meal: it requires careful planning and systematic performance. When you have done your work as a technician, you are ready to face your most important task: reporting. Just when you think you have finished, you have in fact just begun.

When they have completed a technical task, many technicians launch right into writing the report immediately. If they are conscientious, they do so in order to get the job done. If they are like most of us, they do so in order to get it over with. Writing the report right away can be risky, however. If you do not know exactly where you are going, you may end up at the wrong place. You may get tangled up in particulars by not knowing exactly what you are getting into. With the best of intentions, you may fail to complete the job if you have not planned and worked systematically. Planning beats inspiration nine times out of ten.

When you approach writing systematically, you find writing easier and you find yourself doing a professional job. This chapter summarizes some points discussed in earlier chapters and presents a brief guide for preparing your report systematically. The objectives of this chapter are to enable you to:
1. Think through a report before you start to write.
2. Divide your writing tasks into manageable steps.
3. Separate the writing from the editing.
4. Edit efficiently.
5. Produce your report.

A systematic procedure for preparing your report consists of the following steps:
1. Define your purpose in terms of audience.
2. Establish the basic structure.
3. Arrange the discussion.
4. Write the discussion.
5. Prepare the first draft.
6. Prepare the appendices.
7. Edit the prototype draft.
8. Prepare the final draft.

9.2 Defining Your Purpose

This is the first step—and the crucial one. You must define your report purpose so that your report will do exactly what you intend it to do for its audiences. To define your purpose, you prepare a three-step purpose statement, analyze the needs of your primary and secondary audiences, write the subject line and heading, and write a concise purpose statement. Because the report purpose must be understood in terms of the specific

audiences, in this chapter we combine the audience analysis with the purpose statement, even though we earlier discussed the two separately. In actual practice, these two steps are performed simultaneously.

9.2.1 Outlining the Three-Step Purpose Statement

You prepare or at least think through a three-step purpose statement to demonstrate the organization's need for your technical activity. For example, you may have been asked to run a routine test, and you have done so without asking why. Although the tests may have been technically quite routine, the organization's need for them probably was not. You must determine exactly why those tests were needed.

The three-step purpose statement completes your technical activity. When you have finished a test, you have results. How do you interpret those results? Obviously you explain them in light of the organization's needs. For example, your company's expressed concern for water quality will tell you how to formulate your conclusions and recommendations. Once you have determined the BOD (biological oxygen demand) factor, you must realize that concern about water quality was behind your assignment. This leads you to the specific conclusion that the water quality poses no health hazard. You must understand the organization's need before you complete your analysis and make the appropriate judgments.

This will not always be easy. You will have to look beyond your own department, since even your supervisor may be unconcerned about the organization's need. But a little experience in thinking about your work from the organization's viewpoint will make your task easier in subsequent reports.

9.2.2 Analyzing Audience Needs

You analyze your audience needs to make sure your report will tell its readers what they have to know. When you prepare a three-step purpose statement, you also need to identify your primary readers and their needs. Further, you need to consider who else will be concerned about the results, conclusions, and recommendations of your report. Identify all the possible readers, thinking about their interests and concerns, so that the opening segment, the discussion segment, and the various attachments will address the appropriate readers. Ask yourself whether you should obtain any new information or rethink any part of your investigation.

9.2.3 Writing the Subject Line and Heading

When you have clarified your purpose and identified your audiences, write the subject line. This statement will focus your thinking during the preparation of your report. The subject line therefore identifies your topic and states your purpose. When you have formulated the subject line, complete the rest of the headings. Identify the primary readers and yourself, by name and role. Cite any project codes or prior documents that will place your report in context. Put down the date and list the secondary readers who will receive copies of the report.

9.2.4 Writing the Purpose Statement

Next, write a concise purpose statement as the first paragraph of your report. Whether this is one or several sentences, it is a selective condensation of the complete three-step purpose statement. This gets you started on the actual writing of your report.

9.3 Establishing the Basic Structure

To establish your report's basic structure, formulate your conclusions and recommendations. Your purpose statement will tell you how to analyze your technical investigation and interpret its results. When you have formulated the appropriate conclusions and recommendations, you are ready to complete the opening segment of your report, the most general component of the general-to-particular structure.

9.3.1 Formulating the Conclusions and Recommendations

Write out the conclusions and recommendations as a list of very direct statements. For a single-spaced report two-page or less, the conclusions and recommendations usually total no more than fifty or sixty words.

Forget about all the time you have spent and all the work you have done. Your readers want the results of your work, not a history of how hard it was. Keeping your primary audiences and your purpose statement in mind, state your conclusions and recommendations as directly as you can. Do not explain and elaborate here; the discussion segment is the place for that. Many writers present conclusions and recommendations ineffectively because they overwrite them. If you find yourself continually overwriting at this point, you should arrange your discussion—Step 3 of preparing your report—before writing the opening segment—Step 2.

9.3.2 Writing the Conclusions and Recommendations

You already had a list of sentences stating your conclusions and recommendations. Now format these sentences to complete the second unit of the opening segment. Immediately after your first paragraph, the purpose statement, put either the heading "Conclusions and Recommendations" or two headings—"Conclusions" followed by vertical white space and then "Recommendations." After the heading or after each of the two headings, write an introductory sentence, if one is needed, and then present your conclusions or recommendations.

These are best presented as lists of sentences, at times, even as numbered lists. Thinking of them this way will help you keep them short and to the point. Try to maintain grammatical parallelism in your list, because this makes your reader's task easier. Present only the bottom-line results as documentation of your conclusions.

Once you have finished this second step, you have the first draft of the opening segment.

9.4 Arranging the Discussion

Your discussion supports your conclusions and recommendations. If you have successfully completed Steps 1 and 2, you will be able to design a discussion that is more than just a blow-by-blow account of your investigation or technical activity. The discussion should be selective, meeting your audiences' needs and your own purpose.

9.4.1 Choosing an Appropriate Arrangement

Having drafted your conclusions and recommendations, arrange the particulars of the discussion appropriately. Your primary concern is to explain your conclusions and recommendations. Look at the list of eight possible discussion patterns in Chapter 7 and choose one appropriate to the purpose and opening segment of your report. If at all possible, choose a persuasive pattern. When you choose a pattern to meet your purpose, you will focus your discussion on the most important particulars, on the details that support your conclusions and recommendations, rather than on those that merely explain your technical investigation.

Also you must decide what other information to present and where to put it. Does the discussion itself need an introductory paragraph or a summary paragraph? Do you need a unit on equipment or test specifications? If so, should it come before or after the support segment? Most additional information comes immediately after the main discussion.

9.4.2 Selecting the Relevant Information

At this point, you should decide what information can be put into appendices and attachments. Given your purpose, primary audiences, and secondary audiences, decide what information is not absolutely essential for your discussion. Put these in attachments. Do not be reluctant to send some secondary readers to the appendices and attachments, but try to provide in the discussion what your primary audiences need. Remember that your discussion segment should be efficient. Do not overwrite the discussion by stuffing all the information you have into it.

9.5 Writing the Discussion

If you have conscientiously followed this procedure so far, you should be able to write your discussion confidently and effectively, without losing your sense of direction or getting bogged down in particulars. Writing the discussion requires preparing a sentence outline and then writing the separate paragraphs.

9.5.1 Preparing an Outline

You prepare an outline so that you first state a point, then explain or support it. Like the report itself, the units and paragraphs of the discussion should go from the general to the particular. To prepare the sentence outline, follow the segment pattern you have chosen for the discussion. Write out a complete sentence for each item of the pattern. Then write a sentence for each subordinate idea in each step of the pattern that puts a series or group of ideas under that one heading, such as a point-by-point comparison, for instance. Also write out one sentence for each additional unit of information, such as the background or the test specifications, that will come before or after the more important material called for by your outline. These sentences form the skeleton of your discussion.

9.5.2 Writing the Paragraphs

You now are ready to write a complete draft of the discussion, with each sentence of your outline probably forming the core sentence of a paragraph. Explain or support each core sentence according to an appropriate paragraph pattern. Concentrate on the core idea while writing each paragraph so that you do not try to explain two ideas at once or try to cram in too much detail.

As you write the paragraphs, do not overwrite or overexplain. As you go along, follow the contextual editing procedure, keeping the sentences short and direct and the paragraphs short. Not many sentences should be over fifteen or twenty words long, and not many paragraphs over sixty or seventy. You can use these numbers as a rule of thumb: if you find a chain of thirty-word sentences or a two-hundred-word paragraph, the chances are that you have overwritten the discussion. Remember, the key to success in writing your discussion is to take one thing at a time, rather than trying to do it all at once.

9.6 Checking the First Draft

When you have the first draft of your report, you must evaluate it as a whole, format it, anticipate figures and tables, and have it typed.

9.6.1 Evaluating the Whole Report

Read through your report from beginning to end. Does it hang together? For example, the heading must lead into the purpose statement, and the conclusions and recommendations must follow immediately. The discussion must develop inevitably from the conclusions and recommendations. Perhaps the connection between your discussion and the opening segment needs to be clarified. If so, examine the ways suggested in Chapter 4 for connecting these two segments, revising your handwritten draft to make the appropriate connection. Perhaps you need to signal the ending of the discussion. A discussion often ends on a minor particular, giving the report readers a sense of incompleteness. Look at the subject line, then ask yourself whether the discussion really completes your report. If it does not, consider your organizational problem and your report's audiences, and write a brief concluding paragraph to sum up the conclusions or to state what the next course of action should be. A single sentence is often a good way to end a one- or two-page report effectively.

Check whether the report is balanced. Look at the ratio of material in the opening segment compared to that in the discussion segment. The opening segment should be only about a quarter or a third of the total length in a one- or two-page report. The opening segment of longer reports should be no longer than a page (single-spaced format), except in extremely unusual situations, such as when the heading alone is extremely long and complex. Neither should the discussion segment be too long. Now is the time to cut your discussion down to size. The usual length for reports in the standard single-spaced format illustrated in this text is from one to three pages. If your report is longer, perhaps you have not used appendices and attachments effectively.

9.6.2 Formatting the Report

Using appropriate signals for the typist, block out your report according to any standard format specifications. Then add your own formatting to clarify the structure of your report, to make the report easier to read, and to make various parts of the report more accessible to a diverse audience.

Standard formatting, such as that shown in Figs. 4-1 through 4-4, usually establishes the basic parts of the report and specifies the report headings. If your organization does not have its own standard format, impose a standard format on your draft. Format the heading; perhaps signal the conclusions and recommendations or summary information. Use a major heading to set off the discussion from the opening segment. Finally, reinforce these divisions with horizontal and vertical white spacing and with numbering. You should establish a consistent format for similar types of reports. Your typist will become familiar with it, and your audiences will come to rely on it.

In addition to standard formatting, you should format the discussion. Many reports are less effective than they could be, because the discussion is presented as one big block of prose with only paragraph formatting. Introduce subheads to separate different types of supplemental material and to signal the important points of your discussion. Use horizontal and vertical space, as well as numbering, to supplement the paragraph formatting. Additional vertical space usually accompanies subheads and is often used by itself to mark smaller subdivisions in short reports. Again, you should establish a regular format so that your typist becomes familiar with it.

9.6.3 Anticipating Figures and Tables

Now is the time to read through your draft to determine what and where figures and tables are needed. Your concern at this point is to select the most important figures, tables, or both, which belong in the main text itself, not in the appendices.

You may or may not actually prepare your figures and tables at this time. It is important, though, to determine where any visual aids will be needed and what size they will be, so that the typist can leave space for them when typing the draft. You can often prepare the figures and tables while the draft is being typed, or you can wait until the draft is finished and then ask the typist to prepare them. When possible, prepare your main tables and figures before giving the manuscript to the typist. The typist can then determine the spacing and where the figures and tables should be inserted.

9.6.4 Having the Report Typed

Before you send your report to the typist for a

prototype draft, read through the entire manuscript to check format consistency and to do rough paragraph and sentence editing. However, sentences are usually edited best from a typed draft, so make only minor editing changes, to avoid messing up the handwritten draft any more than necessary.

The typed prototype draft lets you see what the report actually will look like, so you get an overview of its proportion and length. Admittedly, the single-spaced format of the prototype makes sentence editing more difficult than it would be with a double-spaced draft. However, the advantages of a prototype draft outweigh the sentence-editing disadvantage. You can edit the sentences of single-spaced text in the margins or by rewriting some paragraphs on a separate sheet of paper.

The typist should prepare a prototype draft with the figures and tables as finished as possible. Freehand sketches should be inserted if the final art-work is not yet finished.

9.7 Preparing the Appendices

While the body of the report is with the typist, you can finish the appendices, attachments, and file materials. Some of these must be prepared for typing or reproduction; others just need to be identified and collated. The appendices pose problems when they are not especially prepared for the report. Although graphs, printouts, and calculations can sometimes be used in their existing form, more often than not, they must be redone. Writers too often fail to be selective in their appendices, confusing the appendix with the file—or the wastebasket. It is inefficient to send readers to an appendix, and it is even worse to present them with undigested data.

The appendix items must be identified and classified into logical groups. If you have several appendices, they should not be arranged haphazardly. Background and supplemental items can be grouped together. Items from your investigation can be grouped together according to the different activities. The various appendices are ordered according to two somewhat conflicting principles. First, they are arranged in the order they are mentioned in the text. Second, they are arranged in decreasing order of importance and according to similarity of subject matter. Sometimes you can resolve conflicts between these two principles simply by changing the references to the appendices in the text. Mention of appendices in the text is often a function of how you want to arrange them. Readers usually do not need to turn to the appendices to understand a point being made in the text. At least, that is true if you have designed the discussion well.

Each appendix item should be prepared separately. Write each item as a distinct segment, introducing only minimal framework and interpretive material for each item, as well as headings. Appendices items are generally presented in telegraphic or outline form and are heavily formatted. These items usually rely on tables and figures to present details, yet they should be able to stand by themselves as independent units. Because appendices are often referred to by themselves and sometimes are even duplicated and distributed separately, they require this individual attention. A good test is to ask yourself whether the appendix material could fall out of the report and still be understood by a passerby who happened to pick it up. If not, you still have some work to do.

Attachments and file items seldom require much individual attention. Attachments are just that—independent items that have been prepared in another context, so they require little work on your part. File items just need to be cleaned up, identified, and categorized, so that they can be easily found and used in the future.

When you have the appendices, attachments, and file items ready, give them to the typist for the final typing or other preparation. You should not have to edit or revise them after they have been typed; do this before giving them to the typist in the first place. The typist may have to spend a great deal of time preparing figures or tables for some appendices, so be sure to allow for this.

9.8 Checking the Prototype Draft

When you receive the prototype of the main text of your report, you edit it while the appendices, attachments, and visuals are being put into final form. You need to double check the structure of the report and its contents; then you need to edit the prose and check the formatting and the visual aids. While you edit the prototype draft, isolate yourself from the telephone, your colleagues, and other distractions, so you can read through the report and analyze it without interruptions. You might try to find a place where you can read the report out loud without disturbing anyone. Reading out loud often helps to uncover awkward phrases and grammatical slips.

9.8.1 Reviewing the Whole Report

Now is the time to review the complete report (except the appendices, of course) to make certain the basic thinking and structure are sound. If there are any problems that cannot be handled by editing, you have to go back to a previous step of the report-preparation procedure; it is of no use to waste time on sentence editing. Once you go back to make a major change, you cannot predict what will happen to the report. What may seem simple at first can lead to extensive rewriting. Thus, you must satisfy yourself that the report is basically sound before you move on to editing it.

9.8.2 Editing the Report

Few of us can edit our own reports easily, so you should not expect to be able to edit yours just by reading it through sentence by sentence. We suggest two techniques. You should edit the paragraphs in context and then edit single sentences in a few selected paragraphs, that is, spot-check the paragraphs.

Because paragraph sentence patterns help determine the style and structure of the single sentences, you should edit your discussion paragraphs first. If your report is long, select three or four paragraphs to edit, following the procedure described in Chapter 7. The purpose of editing at this point is to uncover any serious problems with how each of your paragraphs works as a whole. Editing the paragraphs in the central section of your discussion will alert you to weaknesses in paragraph organization and to some of the individual sentence problems, if there are any. Then, if necessary, you can edit the entire report to deal with those specific problems.

Next, edit the sentences in a 100- to 200-word section of the central segment of your discussion. Edit the sentences of this passage according to the procedure in Chapter 6. The purpose of this is to uncover your own particular sentence problems. Suppose you notice several indirect "it . . . that" constructions and too many flabby sentences with two main clauses connected by "and."

When you edit your entire report, you will know you should look specifically for other "it . . . that" constructions and for flabby sentences, because if you have made certain mistakes in one place, the chances are that you have made the same mistakes elsewhere.

Finally, read through the entire report, out loud, if possible, editing the single sentences as you go. This method, however, will probably not be as productive as spot-checking. You should read through the entire report a second time to catch whatever spelling and punctuation problems you can.

9.8.3 Checking the Formatting and Visual Aids

Now, review each item of the report separately to check the formatting and the visual aids.

To check the format, scan the entire report several times. First, scan it to make sure the headings are consistent, then again to make certain the numbering system is consistent. Finally, scan the entire report to make sure white space is used consistently. You might be surprised how often inconsistencies persist in final drafts.

You should then review your visual aids. First, evaluate each figure and table according to the criteria presented in Chapter 8. Pay particular attention to whether the visual aids are integrated with the text by the references and explanations. Second, determine where additional figures or tables, if any, should replace or complement passages of the text. Up to now, you have spent your time writing. This may be the first time you have examined the text with the visual aids foremost in mind. If any additional table or figure is needed, you can insert it into the text with scissors and tape.

9.9 Preparing the Final Draft

You are now ready to prepare the final draft of your report. You must have the main text retyped, have all appendicies and attachments prepared in final form, have the final draft proofread, and have multiple copies of the report prepared.

9.9.1 Putting the Report Together

The report now is assembled as a whole, a mechanical operation that should not be neglected. This is the first time you have the entire report in your hand at once.

The typist should be able to type the final draft of the report without difficulty. The editing and formatting changes have been made directly on the prototype draft or on separate sheets attached to it. The precise spacing for visual aids, subheadings, headings, and the main heading of the report have been decided.

Making sure the appendices and attachments are in their final form requires careful attention. These have been prepared separately, so you must scan them to check for consistent format and presentation. The pagination, cross-references, and numbering of all visual aids must also be checked before the entire report is assembled. If care is not taken with these mechanical details of the appendices and attachments, the final report may appear to be in bits and pieces when it really is an integrated whole.

You should not review the final draft until it is completely assembled. Your typist should understand what the entire report is supposed to look like and should be responsible for assembling it. The typist is in the best position to do this because he or she will be typing it, preparing the visual aids or having them prepared, collating the appendices, and obtaining copies of the attachments. You usually are involved in other projects by this time and therefore will not be able to keep track of everything.

9.9.2 Proofreading the Report

When you receive the report from the typist for proofreading, consider it as a finished product. Your purpose now is to make sure that it is polished.

First, you should proofread the report yourself. Try to read the report word by word, without getting caught up in the content. You are searching for typographical errors, misspellings, grammatical mistakes, and blatant mechanical errors. Pay particular attention to the main report heading, the headings, the subheadings, and the visual aids. These often contain typing errors that go unnoticed because no one has proofed them carefully. Double check all numbers in the report against your handwritten manuscript and original sources. Look at each number by itself to make certain that it makes sense. At the stage of the rough, handwritten manuscript, you may have introduced an error that did not get caught in your earlier proofing.

You also should have your report proofread by a colleague. Few of us can proofread our own reports without someone else as backup, because we tend to read typos and misspellings without noticing them. Even your typist is too close to the text to catch all of the errors. A colleague can catch these errors because he or she is unfamilar with the text and thus will read attentively word by word. Remember to have your colleague proofread the report at this step rather than edit it. If you want someone to help you edit, you should

have a colleague help you at the prototype-draft stage (Step 7).

9.9.3 Producing the Final Report

When your report has been proofread, have any necessary changes made. These corrections themselves normally can be proofed on the final copy because corrections are not noticeable in multiple offset or photocopies. Being cost-conscious, few companies expect a letter-perfect original, so retyping will not be necessary.

When you are satisfied with the final version, make a photocopy for your supervisor to review. You should not circulate the master of the report before it is reproduced. When your supervisor has reviewed and perhaps initialled the report, have the number of copies made that you intend to distribute. Make sure that you have a copy made for yourself, so that the master can be kept in the file in case further copies are needed.

You might even keep a second copy. Several weeks after the report has been finished and distributed, review that copy and make notes on what you could or should have done differently. Then, when you write your next report, you can use this marked-up copy as a model.

We doubt that any report writer will follow this or any other writing procedure precisely and methodically. We have presented this systematic procedure for preparing your report to alert you to some of the many problems you might have and to give you a feel for writing systematically, so that you can divide your writing tasks into easily manageable steps that can be done one by one. Your effectiveness as a writer depends on your mastery of a

number of different skills and techniques.
Rather than juggling all these simultaneously,
you should develop a method of writing that
will enable you to apply them only a few at a
time.

Exercises

A. Identify some noncommunication task you do reluctantly but must do often or at least periodically:

1. Identify the objectives of the task.
2. List possible means of going about achieving those objectives.
3. Establish a systematic procedure for doing the task, and write it out in a list similar to our list in Section 9.1.
4. Analyze how this procedure differs from the way you usually do the task.
5. Discuss with the class the advantages and disadvantages of working systematically.

B. It is very difficult to formulate conclusions and recommendations directly and to state them first in a report. Think of an occasion when you should have done this but did not, and explain to the class:
1. Why you did not.
2. Why you should have.

Remember that not all communication situations require this approach; therefore, explain:
3. When and why you would not do this.

C. Sometimes we rush too quickly from first draft to final copy. Explain to the class your experience in such a case. What happened? Could you or should you have introduced intermediate steps?

D. Review the role of the typist in the procedure for preparing a report. Have you ever been in a role analogous to that of the typist? What problems did you have?

10.1 Introduction

Gregg, one of the best students in our technical-writing class, had never made an oral presentation. As he walked up to the front of the class, however, we were sure he would be able to carry it off, even though he looked a little nervous. With his B-plus average in college and consistent A or B grades on all his written reports, he seemed like a sure winner.

He started to talk. As he began, he looked uneasily out the window, then down at the sweaty note cards in his hands. He turned the cards nervously, groping for words to get started. His voice was strained and halting.

"I want to talk to you today about my summer intern project. You see the project was . . . the project was. . . . Well, in my project at the foundry . . . well . . . you see . . . well . . . "

Then Gregg stopped. As he put it later, he welded himself shut. He just froze. Meanwhile we sat there, uncomfortable but unable to help Gregg. We wanted him to do a good job, but the best we could do was to look away and be still. In a second, the room was dead quiet, which seemed to make things worse. After a long pause, Gregg looked up and said, "I'm sorry, I forgot what I wanted to say. I just can't do it." And he sat down.

That scene may sound unreal to some of you who are able to get up and speak in front of a group. It may be hard for you to imagine that a good student like Gregg could have such a hard time with a simple oral presentation. But Gregg is a real person, and that scene happened exactly as we have described it; in fact, it happened twice. We gave Gregg a few minutes to collect himself, and, after we assured him that we all understood and that he shouldn't worry about it, he tried again.

Unfortunately Gregg again welded himself shut and he never did finish that oral presentation.

Many people find giving oral presentations frightening and difficult. Nevertheless, much of the technician's work involves oral as well as written communication skills. For precisely that reason, we would like to introduce you to the oral presentation. Our objectives are to help you to:

1. Understand the professional importance of oral communication skills for technicians.
2. Understand the distinctive demands imposed upon you when presenting technical reports orally.
3. Apply some practical tips on presentation techniques that will make your oral reports polished and effective.

In general, it is not difficult to transfer your skills in writing reports to oral presentation, but the differences between the two types of reports merit some consideration.

10.2 The Importance of Oral Communication

Several studies suggest that you will spend at least half of your total communication time in oral communication of one sort or another. For example, Dr. Terry Skelton's survey, which we cited earlier, finds that in all areas of reporting—sending data, reporting on progress, or making suggestions on policy and procedures—you will do so orally slightly more often than in writing. Of course, you will do much of your reporting both ways, because oral communication frequently precedes or accompanies written communication. If someone is anxiously waiting for the results of some tests, for example, you will probably tell that person about the results as soon as they

are available, rather than making him or her wait the several days it will take to write, type, reproduce, and distribute the final written report. Perhaps you will do it by phone, perhaps by a face-to-face conversation, or perhaps by a rather formal presentation. But in addition to writing about your results, you will probably speak about them as well. There are situations in which you will report only orally. In short, more than half of all your reporting will be oral. Your most important oral communications require careful preparation.

The people you speak to have a much clearer idea of you as an individual than the people you write to. The latter see only your words on paper, but when you report orally, you are physically present. Your voice and appearance are your own. Your physical bearing is—or seems to be—you. In the memory of the people you work with, you probably come through much more clearly, as a face and a voice, than you do to your readers, for whom you are just a series of typewritten words on a sheet of paper. When you speak, you never do so anonymously, but your written work will often go out under several signatures in addition to your own or perhaps without your signature at all. Your readers may think of you as an individual only when they read your reports. But when they hear you speak, when they see you up there in front, they will form impressions and make unconscious evaluations that will stick in their minds for a long time.

As an example, just consider Gregg. We can still remember how he looked and sounded, even what his topic was, even though Gregg graduated more than two years and several hundred students ago. By contrast, we cannot remember his written reports—or even their subjects—nearly as distinctly, except that they were always well done.

To speak in public is to be visible as an individual. If clients, colleagues, supervisors, and managers remember you favorably, your chances of advancement are improved.

10.3 Differences Between Written and Oral Reports

There is nothing worse than a written report read aloud to an audience. In addition to the fact that most people do not read very well out loud, the constraints and opportunities of oral presentations are completely different from those of written reports. The two, therefore, must be designed quite differently. Let us begin with a brief look at the differences between a written report and an oral report. There are five areas to consider: the audience, the design signals, the length and complexity, the focus, and the sender—receiver relationship.

10.3.1 Audience

An oral report has a single audience. No matter what the degree of formality of an oral report, its audience is right there in front of you, all at the same time. You can look your listeners in the eye, and they can hear what you have to say only once. By contrast, the audiences for a written report are usually made up of readers in various places, who can spend as much or as little time as they need to understand what you have to say. As we have seen, some readers of written reports are in the primary audiences (those who must act and make decisions based upon what you present); some are in the secondary audiences (those affected by what you present); and some are in the immediate audiences (those who transmit what you present). When designing the written report, you must bear in mind that it is likely to get very different levels of attention from different readers. Therefore,

you give a written report a two-segment structure.

In an oral presentation, all your listeners are there for the same amount of time; they should all find what you say both understandable and useful. You must therefore design a one-segment report. You should not expect some of your listeners just to disappear—even mentally—while you talk to the others; you cannot expect some to put up with technical jargon and terms they do not understand. Rather, you must reach all of them simultaneously, no matter how diverse their interests and capabilites are. Of course, they really are a diverse group much of the time, but you must speak to them all at once.

The first point to consider, then, is that in oral presentations you have to treat your audience as a single unit. You have to present your oral report at a level all of your listeners can understand; this means you have to be very selective in what you say and how you say it. You should present only the key ideas all of your listeners need.

10.3.2 Design Signals

Oral reports have fewer and different design signals than written reports; therefore, you must be sure those you use are explicit and unmistakable. Once you become accustomed to the design signals in written reports, you will use them automatically to clarify how your report is organized. In a written report, you use visual cues, such as headings, white space, and numbering, to supplement the verbal transitions. You can even put a table of contents at the beginning. For example, a heading in a written report can be typed in capital letters, underlined or both. It can be set off with white space, and it can be accompanied by a numbering system.

In an oral presentation, however, your design signals are almost entirely verbal and therefore must be handled skillfully if your audience is to notice them. You use your verbal "signposting," your transitions, your introductions of visual aids, your voice, and your body as signals of the design of your oral report. You must do this so that your listeners understand the signal the first time, because there is no second time.

10.3.3 Length and Complexity

Written reports can be as long and complex as they need to be to do the job. Oral reports can be only as long and complex as the audience will tolerate.

Most of your reports will be relatively short. It is difficult to listen to a speaker for more than a very short time. After ten, or fifteen minutes perhaps, it becomes very hard to pay attention and to follow what is being said. Oral reports must be short enough to maintain the listeners' attention all the way to the end.

The rate of delivery in an oral presentation, the rate at which a person ordinarily talks, is much higher than the rate at which we read a paper or other material aloud.

If you tape-record a speaker for five minutes and later transcribe exactly what he or she said, you may be surprised at how much typewritten material you get. Although a five-minute talk seems quite short, it actually can contain a great deal of information; most oral reports need be no longer than this.

You know from your classes how difficult it is to absorb information presented orally. You can usually grasp three, four, or perhaps five main points in a lecture, but seldom more than that. Remember all those class notes you have

taken? After five or six points, you just can't keep track any longer, even when your grade depends on it. Your oral presentation audience has even less motivation, so your oral reports must be simple as well as short.

10.3.4 Focus

An oral report does not have multiple uses, so it should focus on your primary purpose. The temptation is to make an oral report a synopsis of a written report. This must be resisted. You will be tempted to cram into the oral report all of the subjects covered in the written report, as though all those points were necessary. A written report usually aims to help someone to make a decision or to act, by presenting information which will justify that decision or action. It further provides a record, documentation for future reference, of the basis for the action and decision. An oral report provides no record; it only provides specific information and advocates specific actions.

Consider this example. Suppose you have been working on the preparation of a maintenance manual for the waste-water treatment crew, and you have successfully completed this project. Your written document includes a great deal of information that you could not and would not want to cover in an oral briefing to the same crew. In the manual, you might cover five major topics in addition to the introduction, the conclusions, and attachments:

1. Process Description
2. Detailed Operations and Controls
3. Maintenance
4. Safety
5. Utilities
6. Emergency Response Plans

You will probably divide each of these major topics further. "Process Description," for example, might be divided into twelve subdivisions:

1. Preliminary Treatment
2. Primary Treatment
3. Secondary Treatment
4. Sludge Handling
5. Scum Handling
6. Septic Tank Dumping
7. Sludge Digestion and Drying
8. Disinfection
9. Phosphorous Removal
10. Sampling
11. Water Systems
12. Under Drain System

These in turn could be divided into smaller units. "Primary Treatment" could have six units, for example, "Secondary Treatment" nine, and so on. The written document will have to be long and complex to cover the great deal of very specific information necessary for effective maintenance.

Now suppose that, after having completed this document, you are called on to give an oral presentation to introduce the crew to the maintenance system. Obviously, you cannot hope to cover all the topics covered in the manual. Although you might be tempted to spend, say, two or three minutes on each of the six major topics in the manual, that would be unwise. You could give only an extremely superficial overview of the system, and in the end you would probably confuse your listeners more than help them. They just could not keep all those details in mind.

A better strategy for you in this situation is to consider the precise purpose of your presentation for the particular audience. Are you there to tell shift foremen about how and when the new system will be put in operation?

If so, the content of the manual is not your topic at all. Or are you trying to identify probable trouble spots in the new procedures? If so, again you have a different focus. In another briefing, you could be explaining to an audience of managers how the new procedures were arrived at and trying to convince them that these procedures are reliable and up to date. If so, again a very specific focus emerges, one very different from that of a written report.

As you can see from this example, the written report has multiple purposes and covers a number of points. The oral report, however, cannot accomplish nearly so much. In fact, it is never intended to. Rather, it must define its own specific purpose and maintain a sharp focus if it is to be successful. That purpose, determined by the audience's needs, may require you to spend all your time talking about something that is not discussed at all in the original report.

10.3.5 Sender–Receiver Relationship

You never write a report with the expectation that there will be immediate response from the readers. Indeed, you may never hear from them directly. When you speak, however, your audience is right there in front of you. You can see whether or not they get the message, and, more important for you as a sender, you can get something from them. You can ask for their input on the spot, incorporating it into the presentation.

Consider this example. As we write this section of the text, we are trying to anticipate trouble spots for you students who will give oral presentations in school and later in industry or business. We are trying to predict your problems and questions, but we must do this on the basis of our experience with similar students, not with you actually here in our office. Suppose that instead we got half a dozen of you together in our office and began to talk with you face to face about oral presentations. Then we would have the benefit of your immediate reactions, questions, and suggestions. We could see from your faces whether or not you understood us. More particularly, we could ask whether you had any questions we had not covered or any suggestions we had not made. Right then and there, we could clarify the murky spots and supplement our coverage.

As you can see from this illustration, the relationship between sender and receiver is quite different in written and in oral presentations. Initially at least, writers have a one-way conversation with their audiences. A speaker, however, has a two-way conversation. Even when the listeners only receive information, they participate by paying attention. This is the opportunity the oral report provides. If you are to take the best advantage of their presence, you cannot ignore your audience, as so many speakers do. You cannot talk at them; you must talk with them. This means not only that you must watch your audience carefully to anticipate what they are thinking, you must actively solicit their questions and comments. Furthermore, you must be quick-witted enough to reshape and supplement your presentation according to their input. You cannot just recite your piece and sit down.

Space does not permit us to explore all of the differences in the constraints and opportunities of written and oral reports. It should be evident, though, that the fundamental components of writing and speaking are different. It is important for you to recognize that a good oral presentation is not just a written report delivered orally. It is a distinctive report for which, although you can apply a great deal of what you have already learned in the previous chapters of this book, you must also be alert to different needs. In the remaining sections of this chapter, we will provide a brief set of suggestions for planning and presenting oral reports.

10.4 Planning and Presenting Oral Reports

Effective oral presentations require careful planning and skillful delivery, no matter how large or small the audience or how formal or informal the occasion. However, as we discuss planning, rehearsal, and presentation, we are thinking primarily of a relatively informal oral briefing to a small group of about five to fifteen people. This is the sort of speaking you will probably do most of the time. The basic principles we discuss can also be adapted to many other audiences and occasions.

10.4.1 Planning The Report

Your oral presentation should be clearly divided into three basic parts, presented in such a way that the audience knows immediately when you are moving from one part to the next. These parts are the introduction, the discussion, and the conclusion. You should plan the speech by thinking out and preparing notes on what you will say in each of these three parts, by preparing the visual aids to accompany these parts, and by rehearsing the parts enough so that you can deliver the oral presentation within your allotted time without your planning notes.

10.4.1.1 The Introduction
For many speakers, the worst part is getting started. They hurry themselves or they

nervously rush into the first visual aid before anyone even knows what the presentation is about. This stage of the speech is critical, because here you establish the purpose of the report and structure you will follow. If your listeners are lost at this point, they are likely to remain lost. Accordingly, plan to devote a substantial part of your time to:

1. Provide a context for your talk
2. State your purpose
3. Forecast your structure

Do not begin the discussion until you have fully accomplished these. In a fifteen-minute talk, three or four minutes might be devoted to these three tasks; in a ten-minute talk, perhaps two or three minutes; and in a five-minute talk, at least a minute.

Your oral presentation is one part of a technical and communication context, so your audience needs to be told what that context is. If your talk is related to a study which began the previous December, you need to take your audience back to this starting point. Why was the work done? Who initiated it? Who has participated in it? What is the current status of the work? Your audience may have only a vague idea about the context of your report; if your listeners are to understand your presentation, they must have a solid understanding of what need is behind your technical work and your talk and of how your purpose relates to their interests and needs.

Tell your audience what you hope to accomplish. Say, for example, "I have been asked by Mr. Howard to tell you something about the current status of the fracture tests. My purpose today is to let you know exactly where we are with the study and when we expect it to be completed. At the end of the talk, you should have a good enough idea of

our schedule to begin planning the next stage of the project." Don't hurry this statement of purpose. Give the audience time to hear it; make sure you have provided a clear cue that this is a purpose statement: "My purpose in this talk is. . . ."

Your presentation is designed to cover three, four, or perhaps five main points, as we have already explained. State exactly how many main points you have and what they are. Say, for example, "To make clear to you why we are recommending this particular model of head-liner equipment, I would like to explain its three basic advantages over its competitors. These advantages are, first, its size and weight; second, its cost; and, third, its ease of operation. Let me now explain each of those three points one at a time."

A fourth task can be added to the end of your introductory segment when needed. Define any unfamiliar terms or concepts you must use in the presentation. If you are in doubt about whether your audience will know the term or concept, say something like this: "Now, as I go along in this explanation of our tests, I am going to have to use the term 'rugosity.' This term may not be familiar to all of you, so let me first explain that 'rugosity' means. . . ."

Devote a substantial amount of time at the beginning of every oral presentation to establishing the context, stating the purpose, and explaining the structure of the report. If necessary, also define any unfamiliar or unusual terms or concepts. No matter how informal or brief, your presentation always needs these tasks to be accomplished if it is to be effective.

10.4.1.2 The Discussion
Some speakers immerse their audience in seemingly endless details. To avoid this loss of

direction, you should divide your discussion into separate parts; you should carefully prepare notes on those parts, so that you know not only what you want to say but what you do not; and you should prepare visual aids to clarify your points and help impose order on the details.

Your discussion should be divided into, at most, five main points, no more. Of course, the specific nature of these parts is determined by your overall purpose. The parts should be numbered and explicitly stated: "My first point in favor of the change of suppliers is that we can cut our delivery time by 30 percent if we change. . . . My second point is that, in addition to cutting our delivery time, we can also cut our cost. . . ." And so on.

These main points should be stated in very precise, parallel terms and should be numbered consistently. "The first cause of the accident was. . . . The second cause of the accident was. . . . The third cause of the accident was. . . ." Do not get careless about stating your main points. Remember that whether or not you get anything else clear to your audience, you must get these points clear. If you are successful, your audience should be able to write an accurate summary of your main points from memory after you have finished speaking. Similarly, if you have done this early planning stage correctly, you should be able to write out a set of very basic notes for your presentation, which might look like this:

Purpose: To explain the experimental procedure for calculating the effects of three different aggregate types on the strength of bituminous concrete mixes. To help the Contractor's Association understand how we arrived at the figures for the construction guide.

Structure of the Presentation:
1. The design of the experiment
2. The results of the tests
3. Preparation of the construction tables

Now, if your oral presentation is to be very casual, you need not go much further than this in actually writing out notes. However, for the sort of informal briefing we have in mind, you will probably want to prepare more detailed notes on each part of the presentation.

Each of the basic parts of the presentation now must be broken down into its own subordinate units. To do that, prepare separate notes on each part. Start by subdividing your several topics into subtopics. For example, in the presentation on testing, the section on the design of the experiment could be broken down as follows:

1. The design of the experiment
   1.1. Experimental setup
   1.2. Preparation of the specimens
   1.3. Measurement of the strength of the specimens

Each of these can then be amplified in your notes to any level of detail you feel necessary for the presentation. The notes above could be developed further:

1. The design of the experiment
   1.1 Experimental setup (show Figure 1)
      1.1.1. The molds
      1.1.2. The mold liners
      1.1.3. The vibratory mold mounting
      1.1.4. The mold caps
   1.2. Preparation of specimens (show Figure 2)
      1.2.1. Mixing the aggregates (Photo a)
      1.2.2. Heating the aggregates and the asphalt (Photo b)
      1.2.3. Compacting the specimens in the molds (Photo c)
      1.2.4. Removing the molds (Photo d)
      1.2.5. Curing the specimens (Photo e)

1.3. Measurement of the strength of the
specimens
   1.3.1. Tension test (show Figure 3)
   1.3.2. Compression test (show Figure 4)

Each of these headings could be expanded into sentence form to state the points you plan to make.

Very simple notes should suffice for informal briefings; for more formal occasions, you can go so far as to write out a detailed sentence outline. Sometimes you can even write out a draft of the complete speech. We repeat, however, that none of these notes will be used in the actual presentation of the report; you use them only to think out the appropriate content and arrangement for the presentation. Later on in this chapter, we will explain what kind of notes to use in the actual presentation.

Don't ever memorize the speech. Only two things can happen if you do, and neither is good. Either you will forget and freeze, or you will sound canned, like the tour guide at an amusement park, and you know how tired and monotonous that can be. Never memorize a speech.

10.4.1.3 Visual Aids

As the outlines above indicate, now you must think about your visual aids. "Figure 1" in the outline refers to a large schematic representation of the experimental setup. Figure 2 is a flow diagram explaining how the specimens were processed, and the photos show each stage of this process. As you prepare the presentation, decide precisely what visuals you will need to clarify each step or point. Plan to use them liberally: having too few visual aids can leave some important points unclear; it is almost impossible to have too many. Make your visual aids very large scale to permit your audience to see and read

them easily. They should be bold, simple, and very precisely done. Unclear or inaccurate visual aids create a negative impression. It is important to prepare your aids at this point, rather than at the last minute, so that you can practice using them and can be sure they are all ready when you need them.

You probably will have to prepare your own visual aids, although some companies have their own graphics departments. Rely primarily on poster board (separate pieces or flip charts) or overhead-projector transparencies, because these do not require you to darken the room during your presentation. Poster-board sheets and flip charts are easily prepared in several colors with the felt-tip markers now available. Effective overhead-projector transparencies require some special equipment and skills, so make certain you have access to these before planning to use transparencies. Poster-board visual aids, unlike transparencies and slides, do not require special facilities and are not susceptible to equipment failure. You should avoid handouts if at all possible because these distract the audience's attention from your presentation. You should not use the blackboard, because that interferes with your relationship to your audience.

10.4.1.4 The Conclusion

For the report's conclusion you should prepare a visual aid which summarizes your main points. You may also want to give some thought to questions that are certain to be asked. If you can anticipate these questions, you can plan and even rehearse your answers.

10.4.2 Rehearsing the Report

Good presentations are rehearsed. You cannot know for certain that you will be able to stay within your time limits unless you rehearse your presentation. Neither can you be sure you will be able to manage the visual aids well unless you have actually worked with them before you do the real presentation. Plan to rehearse the presentation thoroughly. That probably means doing at least three rehearsals: first, stagger through; second, run through; third, dress rehearsal.

The first rehearsal is likely to be a bit chaotic, because you will be uncertain of your material and unfamiliar with how the visual aids work when you first try it out. Therefore, just try to go through the presentation and to use the visual aids without worrying too much about getting everything letter-perfect. However, do stand up and deliver the presentation much as you will on the actual occasion. Stop and take notes on any problems you see as you go along. You will probably need to stagger through the report several times.

This run through should be a timed rehearsal, in which you try to do a polished job of delivering the material and using the aids. A day or two after your stagger-through rehearsals, go through the whole presentation without interruption. If possible, have a colleague watch this run through, giving you comments afterward on how it sounds and on points that might be fuzzy despite your planning. Be sure he or she sits in the back of the room and evaluates your visual aids for clarity and visibility. Ask your observer to take notes and go over them with you afterward, to help you correct any problems he or she has spotted. An immediate repeat run through will be useful.

This dress rehearsal is a timed and fully detailed presentation of the final report. If at all possible, you should do it in the location of the final presentation, exactly as you will in the real event. If anything goes wrong along the way, ad-lib your way out of it, but do not stop.

Just keep going until you have given the entire presentation. An observer might give you some good pointers, but you need an observer primarily to give you the sense of talking to a real audience. A dress rehearsal in an empty room is difficult and not very useful. Your dress rehearsal should come several days before the actual presentation. Then, just to keep tuned up, you may want to do additional run–throughs each day until the actual presentation.

10.4.3 Presenting The Report

If you have planned your presentation well, prepared good visual aids, and rehearsed sufficiently, you should not be overly nervous about getting up to talk. The real work of the oral presentation is all behind you before you get up to speak; the actual presentation should even be enjoyable for you. There are, however, a number of mechanical matters to keep in mind as you prepare and present your talk.[1]

10.4.3.1 The Room
You may not always be able to do anything about the room or its arrangement. However, if you can get there ahead of time and are able to do a little preparation, you can avoid some common problems. Begin by clearing the speaker's area of everything not essential to your presentation. Extra chairs, papers, notes, electrical cords, or hardware can only get in your way or distract your audience. Tripping over an extension cord or pushing extraneous material aside during your presentation is not professional, to say the least. The rule of thumb is that if you are not going to use it, get rid of it. Especially be sure that the area is not cluttered with visual aids, blackboard notes, or displays from other presentations or for other purposes. Your audience will spend at least some time looking at these rather than at you.

---

[1]Throughout our discussion of oral presentations, we are indebted to our colleague, Professor T. M. Sawyer. Those interested in further discussion should see T. M. Sawyer,

While you are getting the room ready, check the seating and the audience arrangement. The seating should make it possible for everyone to see you and your visual aids easily. If necessary, you can move chairs around or reposition yourself. Never accept a bad arrangement just because it is there. Make sure you are positioned to interact with your audience effectively. Even with larger audiences and in formal situations, you should not be afraid to ask people to move if you have to. Audiences tend to cluster at the back of the room or near the door to permit a quick escape. If you find that happening as people file in, go back and ask them to come up front. Asked to move, most people probably will be happy to do so, or at least would be embarrassed not to if you take the right approach. Explain that you are afraid they will not be able to see the charts from way back there. By expressing your request as concern for your listeners, you make it hard for them to refuse without being rude.

Finally, make sure the room is well ventilated and comfortable. The body heat of an audience quickly warms up a small room. If there are smokers in the group, the air will quickly become smelly and stale. Accordingly, open the windows, if any, or adjust the heat or air conditioning to get cool, fresh air into the room. If possible, start off with the room a little cooler than the normal comfort range, so that when the audience begins to warm the room, it will not become uncomfortable.

If you can, see to it that smokers and nonsmokers are separated. A note on the door asking smokers to please sit on the left side of the room will do. As the audience comes in, you can speak to each person individually. The smokers generally will not mind this, and the nonsmokers will appreciate your thoughtfulness.

10.4.3.2 Appearance

You should look neat and professional any time you get up to talk. This does not necessarily mean that you have to wear a suit and tie or a dress and high heels, it just means you should look as poised as possible in the situation. Conservative, not flashy, dress is best, and you should of course be well-groomed. Remember that your audience starts reacting to you before you ever say a word.

While we are on the subject of appearance, let us talk about your stance and facial expression, two areas a lot of speakers have trouble with. The general principle is very simple: stand up straight and look at your audience. You cannot sound forceful and energetic if you are slouching around, leaning on a lectern, with your hands in your pockets, your legs crossed, and your head down. Stand up straight, stay in one place, face the audience, and look people squarely in the eye (or right between the eyes). Try not to look bored with your own presentation or with them.

Smile; let your face reflect your enthusiasm for your listeners and your topic. When you are introduced, smile at the person who has introduced you and thank him or her, then smile at your audience and thank them for coming. Look upon them as individual people who are interested in what you have to say and who want some visual assurance that you are interested in them. You can win them over in a minute if you look vigorous and interested; conversely, you can chill them instantly if you look bored, poker-faced, and dull. Remember that they are there because of an interest in what you have to say; they want or need to know what you are going to tell them. Stand up and react to them in as vigorous and friendly a way as possible.

"Preparing and Delivering an Oral Presentation," *Technical Communication,* Vol. 26, No. 1, First Quarter 1979, pp. 4–7.

10.4.3.3 Notes

Many speakers feel naked without stacks of notes and a 200-pound solid-oak lectern to hide behind, but our advice is get rid of both. If you have a lot of notes, you will be tempted to read them, and that is a sure prescription for boredom. You may even get them all mixed up and soon be lost. Meanwhile that 200-pound lectern will serve as a wonderful barrier between you and your audience. You will lean on it, hide behind it, grasp its edges with clenched fingers, none of which does anything for your relationship with your listeners. The solution is to get rid of both the stacks of notes and the lectern.

We suggest you put all of your important notes for the talk on visual aids that you can show to the audience. You can use visual aids to show the audience the structure of the talk and to let them know where you are, and at the same time to remind yourself what you have to say. You can say something like, "I'm going to be talking about three topics. These are . . ." You display your visual aid:

1. Design of the experiment
2. Preliminary results
3. Possible areas for further testing

Then do the same for each major subdivision of your talk. Most of the particulars in any part also can be suggested by a photo, a schematic, a flow chart, a graph, etc. The visual aids will make a stronger impression on your audience than your verbal discussion does and at the same time will help you keep track of where you are. If you feel you must have some notes in addition to these visual aids, limit yourself to one 3×5-inch card. Put that in your pocket to be used, like a parachute, only in an emergency.

This technique of using no notes other than visual aids, plus a "parachute," has been used for a long time by the students of one of our colleagues, with considerable success. Professor Thomas M. Sawyer points out that since audiences can remember only as many as five main points, there is no need for a lot of complicated notes. He further says if you are going to use notes, you might as well show them to the audience in the form of visual aids. This only reinforces what you are saying, helping your audience as well as you to remember. Finally, he says, all speakers get nervous, sweaty hands that naturally tend to play with and mix up notes. One sure cure is to get rid of the notes entirely. To the novice speaker, that may sound suicidal, but it really does work.

10.4.3.4 Gestures and Movement

During an oral presentation, you may be tempted to keep your hands still by grabbing on to something or by putting them behind your back or in your pockets. However, your hands should operate as they do when you ordinarily talk, pointing, gesturing, underscoring what you say. Start out with your hands at your sides and then forget about them. They will take care of themselves if you let them, especially when you start with your visual aids. Just be sure you do not nervously jingle the change in your pockets or play with a pencil, the pointer, or your necktie. Keep your hands empty and let them do the sort of "talking" they do all of the time. Point; count things off on your fingers. In short, use your hands instead of hiding them.

When you move, move appropriately. Your movement should help your presentation. Step back at the end of a section; step forward to begin a new section. Step to one side to call attention to a graph. Say, "As you can see on this graph . . ." and then stride purposefully over to it and point specifically to what you want the audience to see. Movement gives the audience something to look at and underscores your meaning if you use it well. But really use it, don't just shuffle, wander about, or gesture vaguely. Do not do a soft-shoe routine; stay put, with both feet planted on the floor, until there is a reason to move. Then move with vigor and reasonable speed. Movement always conveys meaning; you do not want it to convey the impression of aimlessness, nervousness, or boredom. Rather, it should point up the transitions from and connections of one idea to another. Like white space or numbering in a written report, it should say to the audience, "Pay attention; this is important."

10.4.3.5 Voice

Most of us talk with variety, emphasis, and energy most of the time. In front of an audience, however, many people's voices become monotonous, low, and flat. To keep your audience's interest, speak loudly but slowly, with lots of emphasis. If you talk too low or too fast, you will blur your pronunciation and force your audience to strain to follow you. If you speak with too little energy, you will lull your audience to sleep. Speak to the people in the last seat of the last row, and make sure they hear you. You may feel you are bellowing, but you are just speaking so the whole audience can hear you easily. In a large room, ask someone in the back to signal you if he or she has any trouble hearing at any time during the presentation, and then pay attention to that person.

10.4.3.6 Visual Aids

Many speakers do not use their visual aids effectively. The speaker puts up the aid and then stands beside it absently, just looking at it. Instead of using it to talk to the audience, such a speaker uses it to avoid talking to the audience. Stand beside the visual aid, look at it once to establish what you want to point out,

point, and then look at the audience. Do not look at the aid again unless you must to point out something else on it. For you, one look should suffice.

Point to visual aids with the nearest hand. This avoids a backhanded gesture which invariably turns you away from the audience, forcing you to talk over your shoulder.

Don't prepare your visual aids as you give the talk. That is, do not use a blackboard or sketch pad. Your rate of talking will always exceed your rate of drawing, and you will be forced to turn your attention and probably your body away from the audience. Prepare all your visual aids ahead of time.

If you are using slides or transparencies, have someone in the audience run the machine for you. Your job is to speak, not to operate the hardware. Simply say, "May I have the next transparency, please? Thank you. Now, as you can see here [point] we have forty-three test runs. Of these, as you can see . . ." Someone else can follow your verbal cues to handle everything but the talking. Of course, this means you will have to have everything carefully prepared, labelled, and arranged, which is as it should be. Do what you are supposed to be doing—looking at and speaking to the audience.

10.4.3.7 Timing

Your rehearsals will have established a theoretical time limit for the talk. In practice, though, talks tend to grow in length, because you will always add a little here and there and you will take a little longer than you expected in front of a live audience. Expect to take up about 10 percent more time than you have in rehearsal. That means that when you rehearse your talk, it should be 10% shorter than your actual time limit.

When you are rehearsing, keep track of how much time each of your talk's major points requires. For example, at the end of the first point, you may have spoken for three minutes, at the end of the second, for six minutes, and so on. This time count will show you whether or not you are on schedule and will help you adjust the length if something unanticipated happens. For example, someone in the audience might interrupt you with a question in the middle of the presentation. That happens sometimes, and it may throw your presentation off. There is nothing more embarrassing for a speaker than going over the time limit and being cut off. Use up your time completely or come in under the limit, but never exceed it.

10.4.3.8 Mathematics

Whatever you can say with numbers you can usually say with words first, and you should do so. Most people do not understand mathematics very well and will appreciate having a simple verbal interpretation before or even instead of the mathematical explanation. Similarly, numbers can be simplified or translated into more comprehensible terms. A pie graph is often much more intelligible than a list of percentages; a bar graph will probably be much more effective than a catalogue of numbers. The rule is to use numbers sparingly and as simply as possible and to explain what you are doing in plain English beforehand.

Never make your audience listen to you explain how something is calculated, for instance, unless you are completely certain they all "speak" mathematics fluently. Since that rule eliminates most audiences, we can almost say never do it at all. Never fill the blackboard with figures as you speak with your back to the audience, as some teachers do. It may work in the classroom (we have our doubts), but you should not imitate that procedure. It will make your oral presentations as dull as certain classes.

10.4.3.9 Humor

We have all heard hundreds of televised speeches in which the speaker, a professional performer, delivers jokes written by a team of professional writers. As a result, we somehow feel compelled to put some humor into our own presentations. We try to start with a joke, because all the professionals do it, and we try to "leave them laughing." The problem is that most of the time it does not work. The joke falls flat, leaving both the audience and the speaker uncomfortable, or the joke has nothing to do with the presentation, so it just wastes valuable time. For these reasons, avoid planning any humor for your presentations. If humor unexpectedly presents itself as you go along, use it, but do not plan anything funny or cute. The professionals can make humor work much of the time, but it will usually backfire on you. This is, of course, particularly true if the humor is at all suggestive. *Never* try off-color humor or use even mildly off-color language in front of an audience. You can only be lowered in their eyes if you do.

10.4.3.10 Questions

After most presentations, you should invite the audience to ask questions, saying something like: "That concludes my presentation. If you have any questions, I would be happy to try to answer them." Then wait perhaps ten seconds, because someone has to break the ice. If your purpose is to establish a question-and-answer dialogue, you can prime the audience by introducing a point about which there are questions. If there are no questions, say, "Thank you very much," and sit down.

If there are questions, as there usually are, you want to handle them well. That is a simple matter. Restate each question as it is asked. This serves several purposes. It gives you a chance to rephrase a badly asked question; it gives you a chance to think a bit before you answer; and, most important, it gives the audience a chance to hear the question. Usually the questioner speaks only to you, not to the group as a whole, so many people in the audience do not hear the question. Restating the question invites the whole group to participate; if you just launch into the answer, you are concentrating on only one person in the audience and ignoring the rest.

Make your answers short and to the point. No audience wants to sit through a long, drawn-out question-and-answer period. Restate the questions and answer them briefly. If a long, complex answer is required for any question say, "A short answer to that question is. . . . However, I am afraid that I do not have time now to explain why that is the case. If you would like to come up after the talk, though, I will be happy to discuss it with you in more detail."

You may be asked a question you cannot answer; it happens to all speakers. When it happens to you, never try to fake an answer. That will simply make you look bad. The audience will see that you are trying to squirm out from under the question and will resent your dishonesty in not admitting that you do not know the answer. If you are asked a question you cannot answer, simply say, "I am sorry I do not know the answer to that. It is a good question, though, and I will look into it and let you know later." Then, always follow up on it.

When you feel the question period has gone on long enough, cut it off. Do not let the questions drag on or begin to run back over material already covered. You can stop the questions very easily by saying, "I am sorry we do not have time to deal with all the questions right now. However, if some of you would like to come up afterward I would enjoy talking with you. Thank you all for your attention." That will end the formal session,

permitting those without questions to escape. Those who want to talk can of course come up, and you can continue as long as you want.

10.4.3.11 Beginning and Ending

Speakers often have trouble at the beginning and end of oral presentations. They are nervous and therefore tend to start talking too hastily and to just trail off at the end. Start and stop your presentation very precisely and deliberately; do not let it fade in and fade out.

To begin, acknowledge your introduction: "Thank you, Mr. Leach," then pause. This gives your audience time to shuffle around, cough, and get ready to listen. Look them in the eyes during this pause, and just wait until you have their attention. Then begin: "My topic today concerns the research project we have been carrying out for the department of transportation. It . . . " Speak up, speak slowly, and take your time. Do not let the beginning of the talk get lost in the noise of late arrivals, coughs, and scraping chairs.

Stopping is another matter, especially if you do not use the question-and-answer technique to end your presentation. If you handle the ending wrong, there will be an uneasy moment when you know you are done but the audience does not. They sit there awkwardly, you stand there awkwardly, and nobody does anything. Then you awkwardly mumble something and slink back to your seat, while the audience waits until someone gives the signal that they can bolt for the door. Instead of letting this happen, when you are done, say so. Say, "That concludes my presentation; thank you for coming," and sit down. You have given the signal that everyone, including you, can relax. People know that it is safe to move, cough, and escape. But remember that you must give the signal. This is not a very difficult trick, but it is one that a lot of speakers have never mastered. Simply say you are done when you are done.

Exercises

A. Here are three individual exercises:

1. Prepare notes and visual aids for a ten-minute oral briefing on the subject of giving oral briefings. Use this chapter to develop a talk to be given at a meeting of office or department personnel under your supervision. Tell them what to do when their turn comes to get up to speak.

2. Prepare a set of guidelines for designers of visual aids, making clear how large the print on visual aids must be to be easily visible at different distances. Other features of good visual-aid design should be included in your guidelines as well. Use examples to illustrate each point.

3. Prepare an evaluation sheet to be used to score oral presentations in your own class. This sheet should cover the important features of oral presentations discussed in this chapter, but it should be simple enough to be put onto one side of an 8½ × 11-inch sheet of paper and to be used efficiently by a large group of evaluators in a short time.

B. Here are three class exercises:

1. If an oral briefing or presentation is well planned and well delivered, the audience should be able afterward to restate the purpose of the presentation, to list the major topics covered, and to summarize what was said about each. If the audience cannot do this, the speaker has failed to make himself or herself clear or to impress his or her message on the audience. The audience's success or failure almost always results from the speaker's success or failure.

As a test of your ability to speak clearly to an audience, prepare and rehearse a five-minute oral briefing on the topic of one of your written reports. Prepare notes stating:

a. Your name.
b. The context of your presentation.
c. The purpose of your presentation.
d. The main topics of your presentation, and specifically what you expect your audience to remember about each.

Give these notes to your instructor when you get up to speak. Do not use notes to deliver the presentation. You may and should use visual aids, and you may use one 3 × 5-inch note card as a "parachute" (typed on one side only). Put the note card in your pocket, using it only as a last resort. Deliver the presentation within the specified time limits.

Your instructor will then ask your audience, either the whole class or a random selection of five or so, to write a summary of your presentation. These summaries should include:

a. Your name.
b. The context of your presentation.
c. The purpose of your presentation.
d. The main topics of your presentation and a brief statement of what was important about each.

The audience members will not be permitted to make notes during your presentation.

Your grade on the oral presentation will depend solely upon the degree of accuracy of the audience's summaries. Your own notes will provide the basis for evaluation. The greater the agreement between your notes and the audience summary notes, the higher your grade for the presentation.

To make this task a little tougher, the instructor may wish to add a question of his or her own to supplement the other four. This might be, for example, a question of definition. For example, the instructor might say, "The speaker defined and used the term 'rugosity.' Provide a one-sentence definition of that term."

2. The signals of the structure of an oral presentation are largely verbal. That is, the speaker uses verbal cues such as "my first point," "my second point," to divide the whole presentation into parts. Test whether a speaker provides clear verbal cues. To do this, tape-record several oral presentations, then replay the tape and hunt for any verbal cues. Make a list of precisely what was said at any point to suggest the structure of the presentation. Now rank the presentations, starting with the one with the most verbal structure cues and ending with the one that has the fewest.

3. A good oral presentation should be visually interesting. The speaker should appear professional, look at the audience, appear interested in his or her own topic, move with purpose, and use visual aids effectively. Test whether a speaker is visually interesting in his or her presentation by videotaping a series of oral presentations. Replay the tape, but turn the sound off. Now rank the presentations solely in order of their visual effectiveness.

Afterword

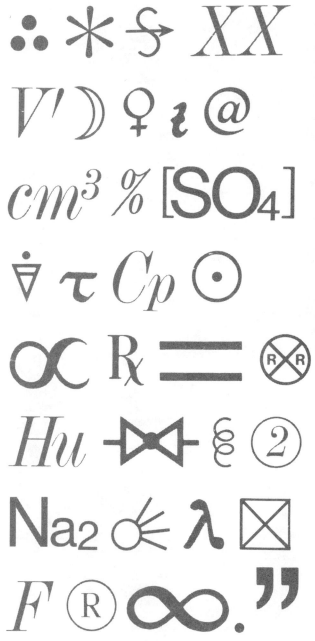

Throughout this book we have operated upon two fundamental beliefs which we hope you have come to understand and share. Perhaps we should return to these two again, for in the final analysis all our suggestions and all your class work in applying them may mean little if you do not begin your career with these beliefs. First, if you are to do well as a technician and are to advance in your responsibilities, you must be able to communicate effectively and efficiently. Second, if you are capable of mastering the complex technical aspects of your role as a technician, you are equally capable of mastering the communication skills necessary to that role.

You must and can master communication skills. If you really believe these two assertions and act on them, you are ready to get on with the business of your career as a technician, paraprofessional, or service specialist.

Index

A

Active verbs, 129–130, 141–142
Amplifying the problem, 75–76
Analysis pattern, 93–100
   example, 96–99
   outline, 100
Analytical paragraph, outline, 163
Appendices, preparing, 201–202. *See also*
       Selectivity
Arrangement. *See* Patterns
Assignment, stating, 27–28
Audiences
   analysis, 10, 13–17, 20
     exercises, 18–20
   characteristics, 15–16
   diversity, 10, 12
   hidden, 12–13
   identifying, 60–61
   managers, 54
   needs, 197
   oral report, 209

B

Background, presenting, 75
Basic structure, 46–69
   designing, 197–198
   examples, 48, 51–53
   exercises, 70–71, 120–126
   negative example, 67–70

C

Causal chain, 106–112
Cause/effect
   paragraph, outline, 163
   pattern, 87–93
     example, 88–90
     outline, 91
Characterizing audiences, 15–16
Clarity, 144–146
   workshop, 147–148
Classifying audiences, 16–17
Clause, 129
Colon, 137
Combining sentences, 134–135
Comma
   fault, 133
   splice, 133

Transmittal, memo of, 101
Typing, 173–177, 200–201, 203–204

V

Verbs, 129
   active and passive, 141–142, 160
Visual aids, 177–189
   captions, 183–189
   effective design, 182–189
   evaluation, 183–189, 203
   examples, 180, 181, 184, 186
   exercises, 191–194

   frequency, 176–181
   interpretation, 183–189
   labeling, 183–199
   oral reports, 215, 219
   placement, 181–182
   simplicity, 182–183

W

White space, 173–175
   examples, 173
Word order, 142–143
Writing skill, importance, 3

Other Books of Interest from Bobbs-Merrill Educational Publishing

Designing Technical Reports: Writing for Audiences in Organizations,
  J. C. Mathes and Dwight W. Stevenson

Word Resources, Thomas E. Walker

Research Papers, William Coyle

Guide to Rapid Revision, Daniel D. Pearlman and Paula R. Pearlman